FERDINAND HAYDEN:
A Young Scientist in the Great West

FERDINAND HAYDEN

A Young Scientist in the Great West
1853-1855

by Fritiof M. Fryxell

Edited by Richard C. Anderson, Phil Salstrom, and Paul Salstrom

Augustana Historical Society
Rock Island, Illinois

Title page photograph: by E. R. Morgan, 1859.
Courtesy National Portrait Gallery.

Augustana Historical Society
Rock Island, Illinois 61201
 Published 2010
ISBN (hardcover): 978-0-910184-47-2

The Augustana Historical Society thanks R. R. Donnelley & Sons Company for permission to reproduce on page 18 the map from *Forty Years a Fur Trader* by Charles Larpenteur (Lakeside Press, 1933); thanks the Gilcrease Museum, Tulsa, Oklahoma, for drawings of Fort Pierre on pages 61 and 62; and thanks The State Historical Society of North Dakota for the image of Fort Union on page 149.

Library of Congress Cataloging-in-Publication Data

Fryxell, Fritiof, 1900-1986.
Ferdinand Hayden: a young scientist in the great West, 1853-1855 /
by Fritiof M. Fryxell; edited by Richard C. Anderson, Phil Salstrom, and Paul Salstrom.
p. cm.
Includes bibliographical references and index.
ISBN 978-0-910184-47-2 (alk. paper)

1. Hayden, F. V. (Ferdinand Vandeveer), 1829-1887--Travel--Badlands (S.D. and Neb.)
2. Hayden, F. V. (Ferdinand Vandeveer), 1829-1887--Travel--Missouri. 3. Geologists--United States--Biography. 4. Badlands (S.D. and Neb.)--Discovery and exploration.
5. Missouri--Discovery and exploration. 6. Geology--United States--History--19th century. 7. Indians of North America--West (U.S.) I. Anderson, Richard C. (Richard Charles), 1930-2009. II. Salstrom, Phil, 1938- III. Salstrom, Paul, 1940- IV. Title.
QE22.H3F79 2010
550.92--dc22
[B]

2009052867

Dedication

To Richard C. Anderson (1930-2009),
meticulous scholar,
patient teacher.
He walked the earth gently.

Contents

Illustrations

Maps

Literary Executor's Preface

Though it might not have been immediately obvious from a visit to his home, Fritiof Fryxell was a tidy person. He kept lists, he organized a large correspondence, he followed systematic schedules for scholarly work. Even his informal exchanges showed this tidiness: the shape of his syntax in conversation and in personal letters was considered and exact. He worked with precision, integrity, and commitment. He finished what he undertook, even when that meant making difficult choices among competing interests.

So it's easy to understand why any unfinished project perturbed Fryxell. When I knew him in his later years, several such works-in-progress covered his dining room table: a biographical sketch of his father; brief histories of the National Association of Geology Teachers and of the geology department at Augustana College, both of which he had founded. These three works we finished together, I editing his drafts and dictated emendations, he meticulously reviewing each version until it satisfied him. When at length he'd conclude, "Well, I think that's as nearly perfect as we can get it," we'd celebrate extravagantly with coffee and Swedish rusks.

But the major project of his last years was a comprehensive biographical study of Ferdinand Vanderveer Hayden, pioneer geologist of the western United States. This work remained in draft form, neat piles of onion-skin typescript which from time to time he would ruefully look over, uncertain how to complete it. The study had been underwritten by major grants from the Guggenheim and the National Science Foundations, whose support naturally increased Fryxell's sense of obligation to finish the work. But in this task, though he tentatively introduced the idea of our collaboration, I felt (rightly) unqualified to assist him.

The present volume helps rectify that problem. Though it doesn't complete the whole project Fryxell and his collaborator Jesse V. Howell had envisioned more than a half-century ago, it retrieves and presents the most finished portions of their study. Professors Richard C. Anderson and Paul Salstrom, along with geologist Phil Salstrom, undertook the task of preparing the manuscript for publication because they believed its material, even in this shortened form, merits the attention of scholars and laypersons alike. And besides its intrinsic interest, this book represents the last work of an internationally important scientist.

Like Professors Anderson and Salstrom, many Augustana graduates remember Fritiof Fryxell as a brilliant and charismatic teacher. He was that. But in addition he achieved international recognition as one of the first major twentieth-century geologists to describe the Teton Mountain range in western Wyoming; as a mountaineer who first ascended and named many of the alpine peaks in the Tetons; as an articulate park ranger (the first in Grand Teton National Park) whose voice and vision helped establish the geographical boundaries of the park. And he was admired as a writer who could describe with scientific clarity and interpret with sensitivity and passion the phenomena of the mountains.

He was also a behind-the-scenes leader in World War II. During that conflict the U.S. government enlisted Fryxell for its Military Geology Unit (MGU), a carefully-selected cadre of scientists who assessed overseas terrain for battle sites and advised military leaders — among them General Dwight D. Eisenhower — on troop movements. Fryxell ended his war career as assistant chief of the MGU.

This brief overview of a full, challenging, and distinguished life helps explain why Fryxell found Ferdinand Hayden particularly fascinating. Fryxell's own geological work in the West drew him to one of the first geologist/naturalists to describe this area. The Hayden Surveys of the years 1867-1879 represent the earliest systematic geological study of the area Fryxell would later deal with in his own scholarly work. Moreover — and perhaps equally important — both men, Hayden and Fryxell, found spiritual nourishment in the sweeping landscapes of the West. Another element linked the two men: both were driven by a hunger to learn that enabled them to transcend economic and social barriers to education. Hayden, whose ne'er-do-weel father ultimately abandoned his family, had to scrimp, wheedle, and sometimes cadge to pay for his schooling. And while Fryxell's privations in no way matched Hayden's, still, as the youngest child of working-class Swedish immigrant parents, Fryxell knew what it meant to put in long days in class and on the job, and often to lunch on a nickel candy bar so that tuition could be managed.

In undertaking this book, Fryxell set out to reveal the youthful explorations of Ferdinand Hayden, a topic that's received scant attention in most works about the period. The epistolary format and Fryxell's gift for clear explanation make complex material accessible and immediate. But the book is more than the personal record of a budding scientist. As Fryxell pointed out in an introductory note (quoted by Anderson and the Salstroms in their Editors' Preface), it's about the world that surrounds Hayden's work. It's about American scientific history, political issues in the American West during the mid-nineteenth century, the inception of the great collaboration between Hayden and Fielding Bradford Meek, and most centrally about the youthful Ferdinand Hayden himself. His extensive correspondence reveals an incandescent personality burning with desire to explore the West and to "advance science," but frustrated by delays in funding, uncertain travel arrangements, inadequate supplies ("I am writing this with a stick," he says at the end of one long letter), and the threatening presence of hostile Native American Indians. Hayden undertook immense tasks with sang-froid. He refers often to the real possibility that he won't return from his western field work, or that he'll come back to "the States" minus a scalp. After recounting one winter foray to collect specimens, when he spent three days lost in a blizzard with no food, he comments wryly: "this was rather hard." Occasionally his single-minded dedication to the advancement of Science, as he styles it, can seem ruthless. He describes an accident which occurred on the Yellowstone River in which "a barrel containing a fine lot of...everything I had" was lost because "the clerk for the Crow Post in passing around the stern by some means fell into the river and in falling caught my barrel and drew everything into the river." He laments the loss of his collection; of the clerk he notes merely "the young man was drowned."

Alternately visionary and practical, aggressive and self-deprecating, eager to learn, warm-hearted and enthusiastic, the youthful Hayden's layered character comes alive here. His personality was counterbalanced by that of his partner and friend Meek, who lacked Hayden's physical and psychological robustness but whose passion for science matched Hayden's own.

Besides the experiences of these two young men, the book offers glimpses into the scientific community of the mid-1800s — into its rivalries, jealousies, and political maneuvering (what else is new?); into the way scientific discovery was carried on and "priority" established; and into the Native American Indians' situation on the Great Plains during that transitional era.

What gives the book its special drive and drama is the format Fryxell chose. Drawing heavily on previously unpublished letters from Hayden, Meek, and others, Fryxell allowed the primary documents themselves to carry the story, supplying necessary background information through narrative links and footnotes. Most of these vivid early letters have not been extensively quoted in print until now. Many of them appear for the first time here, despite the release of several studies dealing with the Hayden Surveys, including two recent biographies of Hayden.

For Augustana College, this volume has special significance. Not only does it represent the last major project from one of Augustana's most distinguished alumni and professors, but it marks an intergenerational collaboration among Fryxell and two of his students: geologist Phil Salstrom, Augustana class of 1960, now retired and living in Wisconsin, and Dr. Anderson, class of 1952, who ended a 39-year tenure at Augustana as Fritiof Fryxell Professor of Geology. Paul Salstrom, who shared in the editing, is an associate professor of history at Saint Mary-of-the-Woods College, Indiana. The Augustana Historical Society is pleased to present this unique collaborative effort.

Ann Boaden, Department of English, Augustana College
Literary executor for Fritiof Fryxell

Editors' Preface

Almost sixty years have passed since Fritiof M. Fryxell began his study of the life of Ferdinand Vandeveer Hayden, pioneer naturalist-geologist, who led early scientific surveys into the Upper Missouri Country. Fryxell's interest in Hayden was probably spurred by his own doctoral geology work in Wyoming, following the trail that the Hayden Survey had blazed seventy-five years earlier. Fryxell grew interested in Ferdinand Hayden's life story, and also in the accomplishments of the survey as a whole, including the work of celebrated members of the survey such as artist Thomas Moran,[1] photographer William Henry Jackson,[2] and geologist Albert Charles Peale,[3] among many others. He visited with Jackson many times and was a close personal friend of Moran's daughters, Ruth B. Moran and Mary Moran Tassin.

Early in his work, Fryxell had assistance from Jesse V. Howell, a successful petroleum geologist who later served as president of the fledgling American Geological Institute. Howell edited that Institute's best-selling *Glossary of Geology and Related Sciences.*[4] The collaboration between Fryxell and Howell to study the life of Ferdinand Hayden began in 1951. Existing records do not reveal how they discovered their mutual interest in Western history and Ferdinand Hayden. Howell made available to Fryxell his extensive collection of items dealing with the history and exploration of the West. Work on the Hayden biography began in earnest in 1954 when Fryxell received a Guggenheim Fellowship to support his research and writing. He was also supported by grants from the National Science Foundation for the academic years 1959-1960 and 1960-1961. By this time Fryxell had decided that this should be a two-volume work — the first volume telling the story of Hayden's early adventures in the Upper Missouri Country and his service as a surgeon for the Union Army in the Civil War, and the second volume recounting his remarkable post-war achievements. By early 1964 the work had advanced to the point where Fryxell could report to Howell that he was "impressed by (1) how far advanced we are with the first volume, (2) how little revision will be required other than condensation (though working up annotation will presents a major problem), and (3) how good our prospects are of completing this volume within coming months."[5] Unfortunately, this assessment proved to be overly optimistic. Advancing age and poor health dogged both men. The death of Howell's wife in 1966 left him a frail and lonely man, unable to help much more with the Hayden project. Likewise, failing eyesight and debilitating health problems made it impossible for Fryxell to carry on. Thus the typescript languished virtually untouched for many years, though consulted from time to time by other Hayden scholars. The "Hayden Collection" that had been assembled by Howell and Fryxell was reposited at the American Heritage Center of the University of Wyoming at Laramie.

Meanwhile, two Hayden biographies have appeared, *Strange Genius* by Mike Foster and *Ferdinand V. Hayden: Entrepreneur of Science* by James G. Cassidy.[6] Reviving Fryxell's biography of Hayden at this late date might seem redundant, but Paul Salstrom, historian and member of the faculty at Saint-Mary-of-the-Woods College in Indiana and a long-time friend of the Fryxell family, recognized the value of the all-but-complete Volume I. He recognized that Hayden's earliest exploits in the Upper Missouri Country as told by Fryxell adds a perspective largely overlooked by Hayden's biographers.[7] Consequently he invited Richard C. Anderson, a geologist who had been both a student and colleague of Fryxell's at Augustana College, to join the effort and finally bring part of the work of Fryxell to fruition. They were joined by Phil Salstrom, another geologist who had studied under Fryxell, who underwrote the cost of producing an electronic version of the typescript and shared the editorial work.

The present volume is not the full "Volume I" envisaged by Fryxell. Rather, it relates events only through 1855, not through the Civil War as was Fryxell's original plan. This volume examines early events and relationships that shaped both the career and the character of Ferdinand Hayden, told largely through the medium of Hayden's own letters. We see his boyhood in a generally dysfunctional family, his education acquired in the face of dire poverty, his insatiable desire to pursue natural history in the field, and his bold temerity when he approached well-known scientists like James Hall and Spencer Baird for their support of expeditions to collect fossils, plants, and animals in the Upper Missouri Country. Of Hayden during this early period, Fryxell said:

> Hayden's work during this period is an epic story which, reconstructed from his personal correspondence (the only source, since his journals have never been found) should rank among the great narratives of exploration. These letters contain the record of Hayden's early interests, purposes and adventures; they reveal his enthusiastic and contagious devotion to science, and the courage, tenacity, and prodigious energy with which he pursued his explorations. Hayden's letters have never been utilized before. [Editors' Note: Foster and Cassidy utilized some of these letters in preparing their Hayden biographies, but they did not treat them at any length, focusing instead on Hayden's post-1855 career.] [These letters] have a spontaneity resulting from the circumstances under which they were written: at the forts on the Upper Missouri, by campfires, on boats plying the Missouri River. (On occasion he

> lacked a pen, and wrote with a sharpened stick. Often his supply of paper was so scant that he crowded every part of a sheet with writing; sometimes he had to write both across the paper and in addition, a second time, lengthwise.)
>
> Dr. Hayden's work during these years included remarkable achievements, giving him the experience that made possible his leadership of the government surveys of later years. Into the seven years preceding the Civil War he crowded an almost fantastic succession of journeys into the Indian country of the Far West, investigating vast areas that, geologically, were then virgin territory. Most of the information set forth is new, of absorbing interest, and highly significant in terms of the history of western exploration and the development of American science.[8]

In an introductory note, Fryxell provided this additional description:

> This book is many things: A life of Hayden, its central figure. A study in personalities — not only Hayden's but of many men whose letters are quoted, for "in their letters the souls of men stand naked." A picture of American science in the 1850s. A previously untold chapter in the early West. An intimate glimpse of a naturalist's ceaseless pursuit of science amidst the tumult and shouting, the delays and frustrations... An intimate glimpse of the workings of the far-reaching hand of the Smithsonian Institution through its assistant secretary S. F. Baird — a glimpse not previously to be had in either the biography of that man or the several histories devoted wholly or in part to the role of the Smithsonian Institution in government and science.
>
> It is a first-hand account, simple and vivid but narrated utterly without embellishment because never intended for publication, of the hazards, difficulties, and frustrations experienced by the scientist who was virtually the first man to undertake systematic field work in the vast Upper Missouri region... It is, in part, the story of deciphering the broad outlines of the geology of the Great Plains and Rocky Mountains. It is the record of the great Meek-Hayden friendship.[9]

Fryxell broached several themes here, and we want to highlight five of them. (1) Not only Hayden himself but other scientists emerge vividly in these pages. (2) Exploration of "the early West" was an important by-product of Hayden's collecting of geological and biological specimens. This required Hayden to operate in tandem with (3) the American Fur Company and its Upper Missouri River steamboats and forts as well as with (4) the U.S. government's Indian Agents. (5) "The great Meek-Hayden friendship" provided opportunities for each to make significant contributions to science that would not have been possible had they worked alone.

A bit of background about each of these five themes will give the reader a useful context. Not only does the youthful Hayden come to life in these pages but also four mature scientists who backed his far-flung efforts. His first patron, James Hall, was one of America's leading geologists and was known for hot-headed rivalries with other scientists. Projecting his massive ego from his geology lab at Albany, New York, James Hall kept American science on tenterhooks.

By contrast, George Engelmann of St. Louis was a mild-mannered and non-contentious botanist. Engelmann provided crucial help there for Hayden's comings and goings on fur-company steamboats that seasonably plied the Missouri River. Another mild-mannered scientist was Joseph Leidy of Philadelphia, who helped finance Hayden's explorations. Like Engelmann and several others, Leidy eagerly awaited the fossil findings and other specimens that Hayden would periodically bring back down the Missouri River. And finally, Spencer Baird of the Smithsonian Institution collected the advance financing that Leidy and others provided and, along with the others, gave Hayden crucial advice about what to look for and how to proceed. At this time in the mid-1850s, Spencer Baird was beginning a long career as the impressario of American science, and he became Hayden's second key patron after James Hall chose to withdraw support from Hayden and nurse his own wounded ego. Lesser-known scientists also enter these pages, but these four were Hayden's mainstays and mentors during his first few years of scientific exploration.

Not just this cast of mature scientists helped to shape Hayden's efforts, but also "the Great West" that Hayden came to love. Hayden and his first-trip partner Fielding Meek were not the very first to bring scientific cogitation into the Upper Missouri Country. Their predecessors' journeys are described early in this book, but a little background might prove useful here about how people traveled up the Missouri River.

Until the U.S. army began sending boatloads of troops up that river in 1855, almost every white visitor got there either by going overland on the Oregon Trail to Fort Laramie and then striking out north on foot or horseback, or else by arranging to ride on one of the Missouri River steamboats of the American Fur Company. Even the official U.S. Indian Agents journeyed up the Missouri on the fur company's steamboats. The American Fur Company had been founded in 1808 by John Jacob Astor of New York City. It wasn't the very first company to send forth from St. Louis those who became known as "mountain men." William Ashley had started doing that in 1822 from St. Louis — but in 1827 the American Fur Company took over and absorbed the Columbia Fur Company, one of several outfits that by then was sending men up the Missouri to trap and trade for furs. In 1832 the American Fur Company founded Fort Pierre (in the center of today's South Dakota) and named that fort for its St. Louis director-of-operations, Pierre Chouteau, Jr. A few years earlier, in 1828, "the Company" (as it was called) had started building Fort Union, located further up the Missouri River where it is joined by the Yellowstone River. Hayden went there on his second journey, and continued 400 miles up the Yellowstone River. Also in these pages he travels much further up the Missouri River itself and brings back many previously unknown fossil species from the Judith River region in today's northern Montana.

The Judith River country was one of the regions in which Indian relations occupied center stage during Hayden's journeys. As you'll see, some Indians were friendly, some were hostile, and many were in-between — their next move unpredictable and perhaps dependent upon circumstances of which Hayden may have been completely unaware. Among the most consistently hostile were the

Brulee Sioux, especially after eighty-six of them were massacred, somewhat arbitrarily, in 1855 at Ash Hollow in Nebraska by U.S. troops under General William S. Harney, nominally in retaliation for the Gratten Massacre of 1854. Thus, the years when the youthful Ferdinand Hayden started his scientific explorations also became the years when the U.S. army first traveled up the Missouri River in force, for the Gratten Massacre provoked Congress to send the army west to control the Sioux. And yet another huge transition also began in 1854 when Congress created Nebraska Territory at the behest of Senator Stephen A. Douglas — a political initiative which started Hayden on his path to governmental sponsorship but which also led directly to the Civil War. Through it all (until the Civil War actually broke out, anyway) Hayden's own personal goal remained the same: to somehow get up the Missouri River year after year with enough money, equipment, and connections to conduct scientific collecting. As the following narrative unfolds, we'll see Hayden's means of pursuing his goal change incrementally before our eyes. After he finally gets back to St. Louis from his long second trip — arriving there half-frozen in January 1856 — he will no longer be living and working from hand to mouth. By that time he had long since lost the help of the volatile James Hall, and even his expedients of gaining sponsorship from one of the leading fur traders and then from the Upper Missouri's U.S. Indian Agent had both worn thin. But now suddenly the U.S. government wanted to hire him and wanted to pay not only his expenses but a salary for him to do exactly what he wanted to do.

One of the most endearing facets of what follows is what Fryxell calls "the great Meek-Hayden friendship." Temperamentally, those two scientists were opposites. Meek was like his name, meek. His patient attention to detail made him a first-rate paleontologist. Hayden by contrast was expansive — omnivorous in his appetite for the unknown. These two young men became lifelong friends and collaborators, one of the most productive partnerships in the history of American science. But first they had to get out from under the dictatorial thumb of James Hall. Facing that task, Hayden could simply in effect say "oh well" and make other arrangements, but Fielding Meek had to wage an inner struggle to finally set himself free from Hall's famous willpower, and that struggle left scars inside Meek, as we'll see.

The editors of this volume by Fritiof Fryxell have not undertaken any original research. Our objective has been to put Fryxell's original typescript into publishable form. Some passages were condensed and a few were deleted to reduce the overall length of the book. The original manuscript lacked fully complete citations of published sources and contained almost no information regarding the location of the letters that it quotes. Fortunately, however, many photocopies and typed transcriptions of those original letters are located in the Hayden Collection at the University of Wyoming, and among the Fryxell Papers at the Thomas Tredway Library at Augustana College. The notes of Hayden's biographers Mike Foster and James G. Cassidy have also been very helpful in assembling the citations for the footnotes.

Finally, Fryxell's research about Ferdinand Hayden was obviously aided by a large number of librarians, historians, antiquarians, and others. But Fryxell had not yet put his acknowledgements into writing. How to track down those contributors at this late date is a task that baffles the present editors. We can only offer our thanks to those persons (known only to themselves) who played vital roles in helping Fryxell assemble the sources from which this work is drawn. And we also gratefully offer our thanks to Kathy Nelson at Augustana College for keyboarding the entire manuscript into an electronic format.

Many other people are also responsible for this book. We are especially grateful to Elizabeth Ducey, Department of Art History, Augustana College, and to Rebecca Mollenhauer, St. Mary-of-the-Woods College, for design and formatting; to Jamie Nelson, head of Special Collections, Thomas Tredway Library, Augustana College, and to Elliott Tinnes, Augustana Class of 2008, for assistance in securing images; and to the many friends and former students of Fritiof M. Fryxell who have encouraged and supported this project.

FERDINAND HAYDEN:
A Young Scientist in the Great West

1

The Boy of the Class

In January 1853, Professor J. P. Kirtland, M.D., of the Cleveland Medical College faculty wrote to his fellow naturalist, Professor Spencer Fullerton Baird, Assistant Secretary of the Smithsonian Institution. He said he was looking forward to Baird's forthcoming work on reptiles, and he mentioned an aspiring young naturalist who was attending his medical lectures in Cleveland:

> A young Gentleman, member of our Med. Class this term, is a thorough Botanist & to some extent familiar with collecting in most departments of Nat'l Science, having been under the tuition as private pupil of Dr. Newberry. He is a graduate of one of our Western Colleges. He is very anxious to spend some two or three years in travelling as a naturalist if he could find income to defray his expenses. Would it be possible to get him attached in any capacity to some of the Foreign expeditions, to the south & western surveys or to any enterprises set on foot by the government? He is trusty & enterprising. Let me know if you please.[1]

Though not named in the letter, Professor Kirtland's young friend was Ferdinand Vandeveer Hayden, then 23 years old and a student at Cleveland Medical College. The "Western College" from which he had graduated was Oberlin. A few weeks later, the young Hayden wrote Spencer Baird directly:

> Cleveland, Feb. 6th, 1853.
>
> Dear Sir:
>
> I am extremely anxious to spend a few years in the study of Natural History. I feel as though I could endure cheerfully any amount of toil, hardship, and self denial provided I could gratify my strong desire to labor in the field as a Naturalist. But I am poor, dependent entirely on my own exertions for my

support, and every longing desire to engage in that most delightful of pursuits to me must be smothered by poverty. I venture thus to address you at the suggestion of Prof. Kirtland who wrote you concerning me some weeks since. He is well acquainted with my circumstances and my wishes and has given me much encouragement. I would ask you what I could do for the Smithsonian Institution the coming season which would give me a support? I am willing to go anywhere, for any length of time and labor with the utmost diligence.

I have taken considerable pains to prepare me for a collector. Could you do anything toward sending me to Oregon or California or any other field that has not been much searched by naturalists? I have been obliged to pay all my expenses through my course of education with the labor of my hands, and consequently but few obstacles would present themselves which I would not cheerfully overcome, and were it necessary to accomplish my purposes I could live as the wild Indians live and endure any amount of exposure and toil without a murmur.

Will you be so kind as to inform me on the receipt of this what you could do for me, and greatly oblige.

Your ob't Servant

F. V. Hayden[2]

Hayden's letter marked the beginning of a voluminous correspondence between himself and Baird. Their close collaboration would continue for over three decades.[3] In this first letter, Hayden reveals a young man who felt a calling and was anxious to prove himself.

Hayden's boyhood and youth had been full of troubles. Born at Westfield in southwestern Massachusetts on September 7, 1829, he was descended from a line of Haydens that went back to a John Hayden, probably from Devonshire, who had arrived in the Massachusetts Bay Colony in 1632 and was among the first settlers of Dorchester, just south of Boston. Ferdinand Hayden's father was a ne'er-do-well named Asa Hayden.[4]

Of his mother's side less is known. His mother Melinda Hawley was born in 1806 and she married Asa Hayden in 1826. Ferdinand's birth in 1829 began their family of three boys and one girl. Apparently they lived most of the time in the Berkshire Mountain town of Lee, Massachusetts, near the New York State line. Ferdinand's father acquired a reputation as shiftless, and served prison sentences in both Massachusetts and New York. When Ferdinand was two years old, the *Rochester* (N.Y.) *Republican* carried this news item about his father:

Asa Hayden, alias Hedding, was arrested Monday evening on board the packet, at or near the lock in Brighton, by constables Steele and Hawley. When arrested he threw a bundle [supposed to be counterfeit bills] through the window of the packet. A woman who was traveling with him immediately threw a bundle of counterfeit notes, described below, into the canal, which was recovered. He was examined before Squire Bishop and the woman accompanying him committed to prison.

> It appeared that he passed to the West a few days ago, having passed several bad notes for tavern bills, stage fare, etc. It also appeared from his papers that he formerly resided in Lee, Mass. and lately kept tavern at Blooming Grove, Orange County...[5]

In 1840 while Asa Hayden served a prison sentence in Massachusetts, Melinda obtained a divorce on grounds of neglect. Leaving Ferdinand with family in Massachusetts, she took her two younger surviving children with her to Rochester, New York, where in 1841 she married a widowed farmer who was eighteen years her senior. As for Asa, after his release from prison he too moved and remarried. He moved to the village of Rochester in northwestern Ohio where he had two married sisters, and there he remarried and fathered two more sons.

Ferdinand was eleven when his mother remarried. Apparently he then lived briefly with her but was soon sent off to Rochester, Ohio to live with one of his father's sisters, Mrs. Horatio Stevens. On a nearby farm lived his father's other sister, a Mrs. Perry. Apparently they looked after him well, and later Mr. and Mrs. Horatio Stevens offered to adopt him — but he declined their offer, not wishing to become a farmer. Locally he became known as studious, and when he was fifteen he walked the sixteen miles to Oberlin and there explained his circumstances to the president of Oberlin College, the famous evangelist Charles G. Finney. Finney enrolled Hayden in Oberlin's Preparatory Department and assigned him work to earn his room and board. After a year's preparation, Hayden entered Oberlin's College Department and graduated four years later, in 1850.[6] The evidence is that, while there, Hayden never received a dime that he did not earn.[7]

Twenty years later, in 1870, Oberlin's Class of 1850 held a reunion. Absent due to organizing his 1870 survey of Montana Territory, Hayden was nonetheless included in the class history prepared by a Rev. Thomas H. Robinson (who at that time was also seeking a position with Hayden's survey):

> Ferdinand V. Hayden was the boy of the Class. He was the youngest and was peculiarly boyish in feelings and manners, susceptible, open to all impressions, frank, unconcealing. With all the trials of poverty and hard study and a misunderstood nature, he yet pushed on, showing an amount of will and persistency that now looks remarkable in the distance. He was a good student, not always thoroughly prepared, yet knowing more always than he could tell. He was enthusiastic, easily carried away, and early showed the bent of his nature in the direction of the Natural Sciences. Ferdinand was little understood. To most of his teachers and classmates he was an enigma. They prophesied no success for him. They deemed him an enthusiastic dreamer who would never conquer in practical life. "Poor Ferd," they would say, "what will become of him?" He seemed never to know what he was going to do with himself. Fond of society, especially of female society, it became sport to play on his susceptibleness. Yet nothing turned him from his course.[8]

Hayden's interest in science made him an anomaly at Oberlin, where most male students prepared for careers in ministry or teaching. But Hayden did find at least one Oberlin mentor, the Professor of Natural History, George N. Allen. The college library holds a sheaf of Hayden's letters to Allen that date from 1859 to 1871. Hayden repeatedly helped Allen obtain scientific specimens, books, and maps; and although Hayden was deluged with applications for positions on his survey, the following May 1870 letter is typical of his ongoing regard for Allen:

> I will always make room for you in such a way that it will suit you. As to your physical ability to endure such a trip, you are, and must be, the best judge. It is quite possible you would go out and enjoy a tour of the kind very much and come back wonderfully delighted and wish to continue them. A Lutheran Minister, Mr. Cyrus Thomas of Illinois, accompanied me last season, and he is now far more anxious to go this summer than ever before, although he leaves a wife and two small children behind.
>
> I did not mean to imply that you could not endure the hardships of such a campaign, for to me the hardships are not great. What hardships there are, the younger members of my party gladly take upon themselves, as standing guard at night, etc. I was merely saying a few words to prepare your mind for rather rough times and rough people. Still you might find in that your greatest benefit, so far as health is concerned, and the greatest pleasure. I am accustomed to look back upon your life at O. as a nice, regular, eminently proper one, while we roam over the limitless plains with a wild joyous freedom from restraint. That is, we cannot take our home-life with us. If I could see you for one hour, I could make myself better understood but probably you catch the idea. Rev. Thomas H. Robinson, of Harrisburg, a classmate, is very anxious to go on the trip. I am afraid I shall be so late in getting into the field that I cannot stop at O. on my way West. Should you conclude to go, I will inform you so that you can meet me at Omaha. I therefore leave the matter entirely with yourself, only stating this, that I shall be glad to have your company and that the door will be kept wide open for you to enter on any terms that may best suit you.[9]

Professor Allen decided not to accompany the 1870 survey, but when Hayden renewed the invitation for 1871 he accepted. He spent the summer in Montana Territory as botanist with the Hayden Survey. (Although Allen's collection of plants was reported on in print by Professor Thomas C. Porter of Lafayette College, a veteran of previous surveys, Allen contributed an interesting paper on "Irrigation methods in Santa Clara Valley, California" to the Annual Report for 1871.) After Allen got back to Oberlin that fall he heard from Hayden, who was writing from Fort Ellis in Montana Territory:

> I was certain you would speak with enthusiasm of this great West. I feel sorry for the man who must leave this world without having his eyes filled with the grand vision. If I have aided you to the sight, I am glad... I am sorry that you were not able to see the wonderful things in the Yellowstone, but when our reports come before the world you will get a pretty clear conception of them. All of my party are well and speak of you with affection.[10]

Hayden evidently had a long memory for those who had helped him. He had been "the boy of the class" but he remembered Professor Allen's kindness, and freely acknowledged it, and reciprocated with a loyalty that continued as long as the professor lived.

2

Recruit for Science

The contrast between the diffident youth seen in previous pages and the energetic, irrepressible enthusiast who now engages our attention is extraordinary. The magic by which this transformation was accomplished may in part elude us, but some of the forces which touched Hayden at this time are evident. They were indeed powerful and vitalizing.

After graduating from Oberlin, Hayden moved to Cleveland, the nearby metropolis on Lake Erie. Perhaps Professor Allen or other Oberlin friends had encouraged him to explore the possibilities of further education. Employment, at least on a part-time basis, was a necessity. Though the specific motives which prompted his move cannot be stated, nor can the exact time be given, the most important fact is that in Cleveland he became acquainted with John Strong Newberry, M.D.

Dr. Newberry, a brilliant young physician and naturalist, had just begun the practice of medicine in Cleveland, and also had turned his attention to local geological studies. By early 1852, Hayden was engaged in the private study of medicine in the offices of Dr. E. Sterling and Dr. Newberry. In return for janitorial and other chores about the offices, he was given bed and room — and instruction, the latter from the junior physician, Dr. Newberry. It appears that, as is so often the case when young men catch visions, Hayden's plans for the future crystallized rapidly through contact with the warm, stimulating personality of his preceptor. In Dr. Newberry, a mere seven years older than himself, Hayden found a friend who manifested genuine concern for his welfare and recognized his latent possibilities. Hayden responded in kind and his life was given new purpose and direction.

Dr. Newberry, like Hayden, had been born in the East, in Connecticut in 1822, and he too had come to northern Ohio as a boy.[1] There his father had opened coal mines near Tallmadge in 1828. The abundant fossil plants in the roof shales of these mines attracted his interest, and by 1841 he had made an excellent collection of "coal measures" plants. In that year Professor James Hall, who was connected with the Geological Survey of New York and was soon to become head of that organization, stayed a few days at the Newberry home and studied the local geology with young John Newberry as his guide. Dr. Newberry's biographer, Professor J. J. Stevenson, was later to assert that "Prof. Hall did much during that journey to secure a right understanding of the relations between the rocks of New York and those of the Mississippi Valley; but he did nothing more important during that trip than that which he did at Cuyahoga Falls in giving a positive impetus to the studies of the boy, whose curiosity had been awakened by the ferns in the coal shales but who now saw equal importance in the molluscan remains of the Waverly."[2] Hall's own biographer, John M. Clarke, wrote, "Newberry was a keen-witted youth, interested in all branches of science, and it seemed as though it were the influence of just such a man as Hall that was needed to guide the young man's activities."[3] From then on, Newberry's determination to be a naturalist was fixed. After the usual preparation, he attended Western Reserve College in Cleveland and received the A.B. in 1846. Continuing his studies at Cleveland Medical College, he became a Doctor of Medicine in 1848, after which he visited Paris for further studies. But, wrote Stevenson, "While he did some work as a medical student he did a great deal of work as a botanical student under Brongniart."[4]

Thus when Newberry returned to Cleveland in 1851, and Hayden a few months later came under his tutelage, Newberry was just on the threshold of his professional career. This was to bring him into the front rank among American scientists — as a physician, particularly through his notable work on the Sanitary Commission during the Civil War, and as a teacher and investigator. After the war and until his death in 1892, he was Professor of Geology and Palaeontology at Columbia School of Mines in New York City.

In writing about Newberry's early work in Cleveland, Stevenson alluded to his influence on Hayden: "Even at that time his enthusiasm proved to be as contagious as it was in later years. F. V. Hayden was studying medicine under him; but soon became so thoroughly a student in geology that the study of medicine was continued only that he might have opportunity to do geological work in the far west; there being no other way than that of securing an appointment as surgeon to some exploring party."[5] So, as James Hall may be said to have "discovered" Newberry, the latter enkindled Hayden, who later in turn provided inspiration and opportunity for numerous other gifted young men.

Though there is little detailed information about this formative period in Hayden's life, it is clear that other men also played important roles in it. Among these was the famous James Hall himself. John M. Clarke, Hall's biographer, tells of this:

> When Hall was returning from his Lake Superior field work in 1851, he stopped at Cleveland to call on his young friend John Newberry and there was introduced to an energetic youth who had been making some geological studies in the neighborhood and who wanted to enter on a career in natural sciences by way of the medical school, as many of the men of that time chose to

James Hall (Courtesy of New York State Museum, Albany, New York)

do. This youth was Ferdinand V. Hayden, in later years to become director of the United States Geological Survey [*sic*]. Hall proposed to Hayden that he enter the Albany Medical College, holding out to him the promise of geological service when his course was over. It was too tempting a proposition to resist and the young man came on, entered the school and lived at Hall's house, meanwhile tramping the classic grounds of the Helderbergs and Schoharie [a celebrated mountain and creek in the Catskill Mountains of upstate New York]. As Mr. [Fielding Bradford] Meek was then in Albany, it was at Professor Hall's table that these two men came into a first contact which led to a long standing scientific partnership.[6]

However, Hayden did not at once move to Albany after meeting Hall; as a matter of fact, he did not actually take up residence in Albany until September 1853. Meanwhile he remained in Cleveland where his private study with Newberry, begun in early 1852, was continued into 1853.[7] And while at Cleveland, he began his more formal medical education by attending a term of lectures at Cleveland Medical College (at that time the Medical Department of Western Reserve College). Whether this term of lectures was in the fall of 1852 or the spring of 1853 is not clear; the College published no catalog for that academic year, and alumni records give no information except that Hayden was a member of the Class of 1854.

Hayden's attendance at Cleveland Medical College brought him into the classroom of Jared Potter Kirtland, M. D., Professor of Physical Diagnosis and the Theory and Practice of Physics. Dr. Kirtland was a successful practitioner and professor (he served on the faculty from 1843 to 1863), and a naturalist of wide interests, a student of "fishes, reptiles, shells, animals, and geology," especially "land and fresh water shells." Dr. Kirtland had inspired the formation of the Cleveland Academy of Science in 1845, and was its only president until 1865, at which time the Academy was reorganized and renamed, in his honor, the Kirtland Society of Natural History. By all accounts, Dr. Kirtland was a leader among Ohio scientists, and a distinguished scholar and teacher. Among his students who achieved fame as naturalists were Robert Kennicott and John Strong Newberry; to this group was now added F. V. Hayden. As has already been noted, it was Dr. Kirtland who brought Hayden to the attention of Spencer F. Baird of the Smithsonian Institution. In later years Hayden often acknowledged his great indebtedness to Dr. Kirtland.[8]

Although Hayden remained in Cleveland longer than suggested by Clarke, it is likely that during his Cleveland sojourn he visited Albany, doubtless for further consultation with James Hall, to examine Hall's great collection of fossils, and possibly also to inquire about completing his medical course at Albany Medical College in line with Hall's suggestion. Hall, who was in the early stages of work on what developed into his monumental "Palaeontology of New York," in 1852 had moved into the one-story red brick "retreat" he had built for himself on the outskirts of the city, on the banks of the Beaverskill River. (Hall's property is now a part of Albany's Lincoln Park, and this building, appropriately marked with a tablet setting forth its significance in the history of American science, is still preserved.) The new structure, familiarly called "the Office," had one large workroom, with galleries to accommodate the thousands of drawers in which Hall kept his specimens, and there was also a room which held his library. In "the Office" Hall worked, and slept, for many years, as did also his assistants, of whom F. B. Meek and later F. V. Hayden were the earliest in a long and illustrious succession extending through nearly half a century.[9]

It was doubtless on his visits to Albany in 1852 that Hayden first met Meek. Since the fortunes of these two men were to become inextricably linked, a further word of introduction is in order concerning James Hall's first assistant.[10] F. B. Meek was born at Madison, Indiana, in 1817, of Irish parentage. His father died when the boy was but three, and Meek's effort to gain an education and later, a foothold in paleontology, proved long and difficult, the more so because of frail health. Eventually he attained his desire when he was appointed an assistant to Dr. David Dale Owen on the geological survey of Iowa, Wisconsin, and Minnesota in 1848-1849.[11] This apprenticeship gave him opportunity to demonstrate his unusual capacity for keen observation and precise description, and brought him to the favorable notice of James Hall, who was seeking an assistant with these qualifications, and one with artistic talent who could prepare illustrations for geological reports. In December 1851, Hall offered Meek a position, which the latter promptly accepted.[12] Meek entered upon his new duties at Albany in January 1852.

The position as Hall's assistant gave Meek invaluable experience and prestige, and thus was a marked step forward in his career, but in certain respects it did not prove a happy alliance. In addition to the fact that he was not robust physically, Meek was ill-suited temperamentally for intimate association with a man like James Hall, who was, according to his biographer, "the personification of superabounding physical vigor" and very much of a "scientific dictator."[13]

Having traced the train of circumstances by which F. V. Hayden, an "obscure student from an obscure institution," came to the attention of scientists of such distinction and influence as Dr. Newberry, Dr. Kirtland, and James Hall (the latter indisputably the foremost American figure of the century in invertebrate paleontology), and also met his future collaborator F. B. Meek, we can next consider an event which concerned all of these men and proved to be of transcendent importance to Hayden and Meek. This event took place in the winter of 1852-1853 and for a brief account we can again turn to Clarke:

> Doctor John Evans, who had gone under Government auspices on an exploring expedition up the Missouri river, returned with a store of fossil treasures from the Mauvaises Terres of the White River region, then in

> Nebraska Territory, bringing unheard of vertebrate remains and extraordinary invertebrates.[14] Professor Hall was at once on the qui vive over this new-found storehouse, almost the first hint of the vast buried treasure of the great West. We may imagine the fever of interest these discoveries excited around the family table at the old house on the corner of Morton Street and Delaware Avenue, and may well believe that Hall yielded not unwillingly to the pleadings of these young men to send them out to that country — a country farther away from Albany then than the heart of Tibet is today; a journey by uncertain boats up the Missouri among still untamed tribes of Red Skins. But the booty was irresistable and after some correspondence with Dr. [Joseph] Leidy [of the Academy of Natural Sciences of Philadelphia] about the chance of being reimbursed for any vertebrate fossils obtained, Hall finally agreed to finance Hayden, though he did not feel that he could release Mr. Meek from the contract under which his work was being carried on.
>
> Hayden's boyish enthusiasm is full of ambitions and hopes. With the opening of spring he starts for Cleveland where Doctor Newberry helps him with his preparations, and from there he writes of his intention to stay all winter in the Bad Lands. Mr. Meek, left behind in Albany, teased to be permitted to go along and though Hall realized the condition of Meek's health and needed his assistance in his New York work, yet he gave way, and so the two young men bring up in St. Louis... Then their troubles began.[15]

It was the appearance of Dr. Evans' published report that triggered James Hall's decision to send Hayden and Meek to the Badlands, since it set forth detailed and incontrovertible evidence relative to "the vast buried treasure of the great West." But previous to this, information about the fossils of the Badlands had been trickling out of the West for a decade and tantalizing men of science. Earlier discoveries had led up to Dr. Evans' work, which in turn was responsible for the initial exploration by Hayden and Meek.

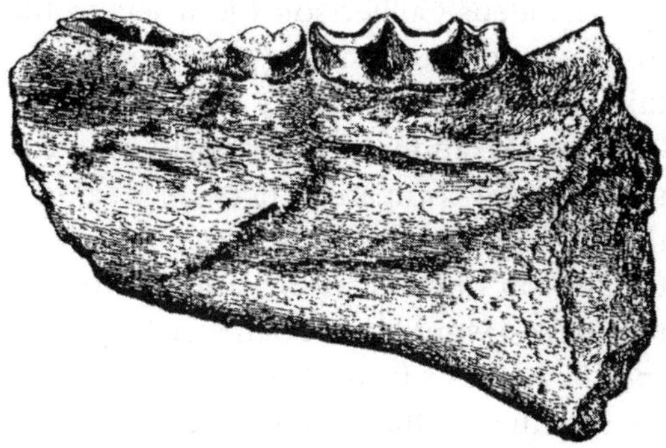

Fragment of the lower jaw of a paleotherium, now known as a titanothere, as illustrated by Hiram A. Prout in 1847. (*American Journal of Science*)

The earliest known collecting in the White River badlands of Nebraska Territory took place in the fall of 1843, when Alexander Culbertson,[16] a prominent fur company official and a man of active curiosity and remarkably keen intelligence, while traveling from Fort Pierre to Fort Laramie, passed through the Badlands and brought out with him some of the fossils he found there. Evidently he submitted one of these specimens to Dr. Hiram A. Prout, a St. Louis physician, who contributed a letter about it, dated December 10, 1846, to the *American Journal of Science.*[17] In 1847 Dr. Prout described in the same publication a fragment of the huge lower jaw of a Titanothere, which he referred to as a Paleotherium.[18] This paper contained the first published illustration of a fossil from the area.

In the summer of 1845, Alexander Culbertson visited his family in Pennsylvania and probably at this time he gave his father, Joseph Culbertson of Chambersburg, some "fossil bones of a new genus of extinct Ruminants, consisting of the cranium and parts of a humerus, ulna, and radius," as they were described when placed on deposit with the Academy of Natural Sciences of Philadelphia in the fall of 1847.[19] Additional fossil bones were deposited with the Academy in 1848. Joseph Leidy described these fossils in eight brief papers prior to publishing a more complete treatment in 1853 in which he set forth the circumstances under which Alexander Culbertson had made his collections and presented them to the Academy.[20]

These finds, though meager, served to arouse widespread interest, and prompted the investigation by Dr. Evans previously mentioned. In 1849, Dr. David Dale Owen was instructed by the Treasury Department, under immediate supervision of the General Land Commissioner, to make a survey of the Northwest Territory, embracing chiefly Wisconsin, Iowa, and portions of Minnesota. He seized the opportunity to extend his area of study farther westward than originally intended, and accordingly in the summer of 1849 sent his assistant, Dr. John Evans, into the Badlands.[21] The preparation and publication of Dr. Owen's report required considerable time; consequently, before the discoveries of Dr. Evans became generally known, the results of another expedition to the Badlands were made public. This trip was made by Thaddeus A. Culbertson, the 27-year-old younger brother of Alexander and a divinity student at Princeton University.

Thaddeus Culbertson's journey, in the spring and summer of 1850, resulted from several circumstances. Investigation of the Badlands and their fossils had by this time become a family interest with the Culbertsons. Thaddeus's influential brother, Alexander, was in a position to guarantee him the freedom of the country, the aid of fur company officials, the service of experienced guides, and payment of the major part of his expenses. The most telling factor, undoubtedly, was young Culbertson's health, as he suffered from tuberculosis and it was hoped that a trip to the West would prove beneficial.[22] On his departure, Thaddeus was charged with specific requests from Professor Baird concerning specimens and information desired by the Smithsonian Institution, and from this organization he received a sum of "about $200" to help defray his expenses. Despite his frail health, Thaddeus Culbertson brought back not only plant and animal specimens but also fossils, termed by Baird "an exceedingly interesting series of Mammalian and Reptilian species including many that had never been described."[23] The benefits to young Culbertson's health, however, proved only temporary, and a few weeks after his return to civilization he succumbed to his disease. The collections he had made were deposited with the Smithsonian, and about two-thirds of his journal, somewhat cut and revised, was published in 1851 in that year's *Annual Report of the Smithsonian Institution*,[24] whereby it received wide distribution. Hayden must cer-

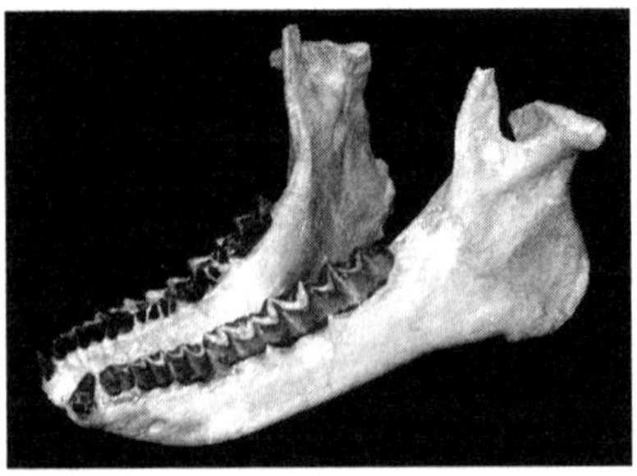

Complete lower jaw of a titanothere. Specimen is about 0.75 m (2.5 ft) in length. (Fryxell Geology Museum, Augustana College)

tainly have read this journal, and probably Meek as well. Undoubtedly it contributed to their desire and ultimately their determination to explore the region whose riches had been sampled by this ill-fated traveler.

Then, late in 1852, appeared David Dale Owen's *Report of a Geological Survey of Wisconsin, Iowa & Minnesota and incidentally of a portion of Nebraska Territory.* The long awaited report consisted of a quarto volume of 628 pages of text, and an accompanying volume of maps and illustrations. The figures of fossil remains, 15 plates in all, were metal-ruled on steel, the first illustrations of the kind produced in America.[25] Dr. Owen himself reported on the Cretaceous invertebrate fossils which Dr. Evans had collected in the Upper Missouri region,[26] Dr. Leidy wrote on the "Remains of Extinct Mammalia and Chelonia" which Dr. Evans had brought back from the Badlands,[27] and Dr. Evans presented, through Dr. Owen, the first "authentic description of the nature of the badland country."[28] As this report contained the first systematic account published on the fossils of the Badlands, it has been considered as marking the beginning in America of studies in vertebrate paleontology.[29]

Dr. Owen's report also included a "Map showing the position of the Bad Lands or Mauvaises Terres of Nebraska from a draft by John Evans of the U. S. Geological Corps" and a view of the Badlands, sketched in the field in 1849 by John Evans, the earliest published illustration of the region. But more graphic than his sketch was Dr. Evans' word picture of the Badlands. He asserted these to be

> one of the most extraordinary and picturesque sights that can be found in the whole Missouri country... From the uniform, monotonous, open prairie, the traveller suddenly descends, one or two hundred feet, into a valley that looks as if it had sunk away from the surrounding world; leaving standing all over it, thousands of abrupt, irregular prismatic, and columnar masses, frequently capped with irregular pyramids, and stretching up to a height of from one to two hundred feet, or more.
>
> So thickly are these natural towers studded over the surface of this extraordinary region, that the traveller threads his way through deep, confined, labyrinthine passages, not unlike the narrow, irregular streets and lanes of some quaint old town of the European Continent. Viewed in the distance, indeed, these rocky piles, in their endless succession, assume the appearance of massive, artificial structures, decked out with all the accessories of buttress and turret, arched doorway and clustered shaft, pinnacle, and finial, and tapering spire.
>
> One might almost imagine oneself approaching some magnificent city of the dead, where the labour and the genius of forgotten nations had left behind them a multitude of monuments of art and skill.
>
> On descending from the heights, however, and proceeding to thread this vast labyrinth, and inspect, in detail, its deep, intricate recesses, the realities of the scene soon dissipate the delusions of the distance. The castellated forms which fancy had conjured up have vanished; and around one, on every side, is bleak and barren desolation.

> Then, too, if the exploration be made in midsummer, the scorching rays of the sun, pouring down in the hundred defiles that conduct the wayfarer through this pathless waste, are reflected back from the white or ash-coloured walls that rise around, unmitigated by a breath of air, or the shelter of a solitary shrub.
>
> The drooping spirits of the scorched geologist are not permitted, however, to flag. The fossil treasures of the way, well repay its sultriness and fatigue. At every step, objects of the highest interest present themselves. Embedded in the debris, lie strewn in the greatest profusion, organic relics of extinct animals. All speak of a vast fresh-water deposit of the early Tertiary period, and disclose the former existence of most remarkable races, that roamed about in bygone ages high up in the Valley of the Missouri, towards the sources of its western tributaries; where now pastures the big-horned *Ovis montana*, the shaggy buffalo or American bison, and the elegant and slenderly-constructed antelope.[30]

As Clarke observed, one can well believe that passages like these aroused a fever of excitement in the young men Hayden and Meek, especially in the impressionable and impetuous Hayden, who was already straining at the leash, chafing "to go anywhere, for any length of time." Nor is it very difficult to imagine something of James Hall's reactions, particularly to that part of the report devoted to the description of invertebrate forms of life previously unknown to science. This was a serious invasion of the province which Hall considered his own, providing competition he could hardly leave unchallenged. The inescapable implication, that even richer harvests were yet to be garnered in the Upper Missouri region, especially in the Badlands, was not to be ignored, and ultimately it broke down his resistance. Time was, it appeared, of the essence: Who would get there first? And whom could he send? Hayden had the answers to these questions: "Here am I. Send me!"

Thus it came about that the dreams of two eager but not very experienced young men became an exciting reality, and preparations for an expedition to the far-off fabulous Mauvaises Terres got underway, albeit gradually and not without delays, misgivings, and, at the last, unforeseen complications which all but wrecked the seemingly well-laid plans.

3

Hayden is Hired

Despite the urgency of the challenge, James Hall did not rush headlong into the Badlands venture, especially since the responsibility for carrying it out had to be entrusted (or so it seemed) to the youthful Ferdinand Hadyen. Indeed, Hall himself lacked first-hand experience with the special problems involved in planning and executing Western explorations. His own field work, though extensive, had been confined mainly to the East and Middle West, where travel and living conditions presented relatively few serious difficulties or frontier problems, such as Indian hostility.

Under the circumstances, therefore, his caution in seeking the counsel of others before making a final commitment is quite understandable. But so is the anxiety of Hayden, who fretted through the early months of 1853 while Hall was soliciting information and advice.

Progress in the planning started slowly, but by spring it developed precipitous speed. Letter exchanges show the stages by which James Hall decided to indeed sponsor an expedition, then modified that decision, but finally prepared Hayden and Meek for their task. The letters illustrate what required attention in planning a scientific expedition to the Great Plains in 1853, in this case an expedition undertaken privately, without the usual advantages of military or other governmental sponsorship. The letters also illuminate the contemporary situation in American scientific circles, showing 1850s scientists with all their human frailties and virtues. Throughout the letters race cross-currents of partisanship and hostility, professional ambition, rivalry, and jealousy. We see a complex interplay of personalities involving not only the principals, Hall, Hayden, and Meek, but also others who became involved in the situation. The preliminaries to the 1853 Meek-Hayden expedition to the Badlands give us a look at the era's science and exploration, including some aspects of 1850s science that, however serious they seemed at the time, now appear ludicrous.

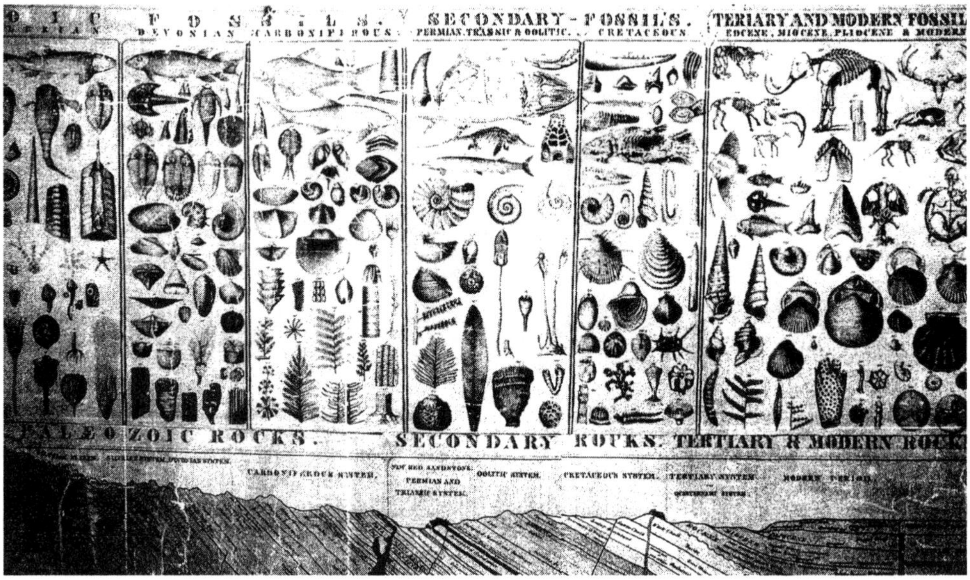

A portion of James Hall's 1853 Geological Chart. (New York State Archives, Albany, New York)

The earliest specific reference to the possibility of an expedition to the Badlands is found in a letter Hayden wrote in response to one of Hall's. Hayden is intrigued, and he tries to impress Hall with his zeal for learning geology:

Cleveland, Jan. 29th, 1853.

Prof. Hall,

Dear Sir,

I received your kind note with great pleasure I can assure you. I think now that I can see my way quite clearly for the coming year. If my health is spared I shall labor with all diligence in any field you may see fit to send me [to]. You spoke in your letter of making some arrangement so that I could remain at Albany when not occupied in the field, if I wished. I had not thought of the possibility of such a thing. Could such an arrangement be made nothing certainly would give me greater pleasure and I should feel under still greater obligations to you, as I am already for the kindness, for the interest and encouragement which you have given me. My facilities in any pursuit, I think, would be far greater there than here. I am somewhat advanced in my Medical Studies, so that, I think, one course of Lectures more would enable me to graduate without much previous study. That course I do not care about taking until a year from next winter, provided I can find employment in scientific pursuits.

The "Mauvaises Terres" expedition I now consider not only possible but probable. But the query with me is, when had I better go? I would like to have your advice. Would it not be better for me to labor nearer home next season and go to "Mauvaises" next Spring after having acquired more experience in field investigations? Would anyone be likely to anticipate me if I waited another year? I would like to become more familiar with Geology as taught in the field before I go, so as to explore the country more thoroughly. However, I will do as you think best.

Lectures close the 23d Feb. which will soon come, and then I shall be ready to go into the field if need be. If I remain in this region next season I would like to have you write me at your earliest convenience and mark out the course you would like to have me take, the kind of fossils you prefer, etc. Could I not labor with you in the N. York field for a month or two this spring? I would then be better prepared to take notes in other states... I could make a collection of Coal Plants for you, of some fifty or a hundred species if you wished them and send or bring them to you. With the instruction of Dr. N. I am making some progress in fossil Botany, also Geology. I can readily determine the genera and most of the species in this vicinity. I spend from 4 to 6 hours per day in the study of Geology. I think I have thoroughly mastered several elementary works... Col. Whittlesey[1] kindly lent me that excellent chart of yours which makes Geology much clearer and more easy of comprehension. I think by Spring I shall be able to take notes with some degree of accuracy.

Prof. Kirtland received a letter a few days ago from Mr. Baird. He says he knows of no situation at present which he thinks would suit me, but inquires if I would go on an expedition if my expenses were paid simply. Whether that would lead to anything better I do not know. I have thought nothing more about it, prefering to accept your proposals.

I see by the papers that the Senate of Missouri have voted a Survey of that State two to one.

I should like to hear from you again as soon as you think best and ascertain something more definite for I can assure [you] I have an ardent desire to get into the field. For some months past I have studied inclusive of lectures from 12 to 15 hours per day and my health is not quite as good as in the Summer. A few weeks in the field would make it as good as ever.

Please reply to some of my inquiries as soon as convenient, and believe me,

Yours with sincere regards,

F. V. Hayden[2]

Following receipt of Hayden's letter, James Hall wasted no time before writing to Dr. George Engelmann, eminent St. Louis physician and botanist, for information about how to mount a Western expedition. Hall's letter to Engelmann follows:

Albany, Feb'y 5th, 1853.

Dr. Engelmann,

Dear Sir,

A young man who is a good Botanist, for a beginner, and who wishes to make geology and palaeontology objects of study collection is desirous to find the means of getting to the Mauvaises Terres. I wish to ask you if you will be so kind as to inform me of the time when the Fur Company's boat leaves St. Louis, and also the names of the persons there in authority to whom

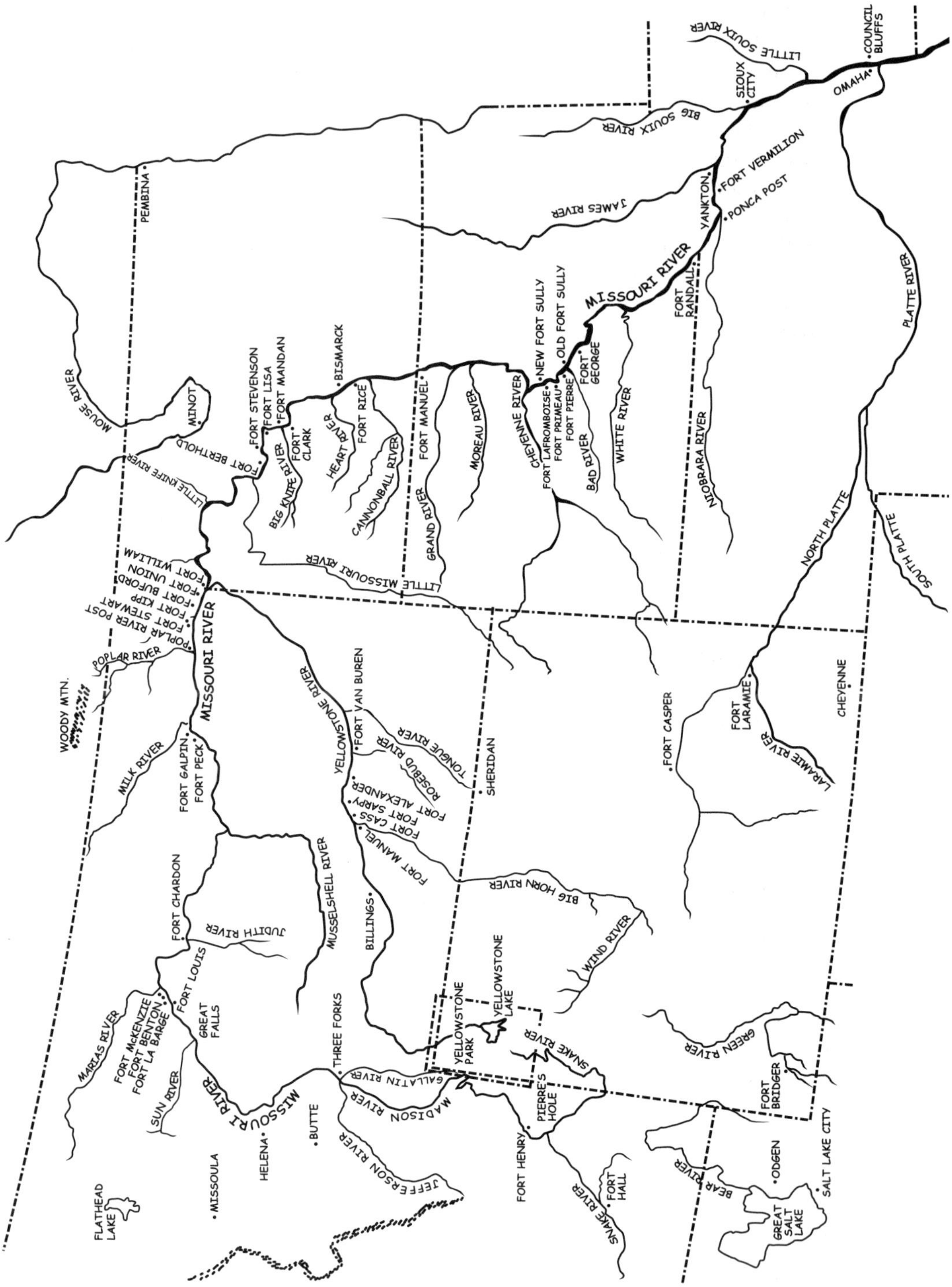

Map based on one of early fur trader Charles Lanpenteur, who mapped the Nebraska Territory. Subsequent state boundaries are noted to provide location framework. (Lakeside Press)

I should apply for the privilege of having this young man go on the boat. I can procure letters from many of our distinguished men here if that will aid the matter at all. You will confer a great favor if you will be so kind as to suggest the best mode of proceeding and inform me at once whether there are any persons here or in New York whose letters or influence will aid me.

I would also like to know whether there will be a second opportunity of going up the River during the season. I presume, however, that both for the objects of Botany and fossils the earliest opportunity will be the best, and this will give him some two or three months to remain in the region before returning.

He will desire to remain long enough to make full collections, but not to delay longer than is necessary his return.

Should you find it convenient to answer my inquiries or refer me to any person who will aid this object you will greatly oblige me.

After my most positive determination to see the iron mountain I have failed to do so. On my arrival at St. Genevieve we had a storm of wind & snow, and my diarrhaea continued so badly that I was prevailed upon to abandon my journey and continue on the boat. Perhaps it was well that I did so for I was much worse for the next day or two and began to be a little anxious to get home. After all, however, I was able to do some work in southern Ohio.

There is a prospect of a meeting at Cleveland in June. I hope you will go by all means. I learn that one branch of your Legislature has passed a bill authorizing a Geological Survey. Can you tell me who will be the man to have charge of it?

I am with much esteem

Very sincerely yours, etc.

James Hall[3]

To these inquiries from James Hall, Engelmann replied in detail:

St. Louis, Febr. 25th, 1853.

Prof. J. Hall,

Dear Sir,

I have tried to get all the information I could for your young friend, and such as it is it is at your service.

The fur company's boat leaves in June as before that time not sufficient water can be found on the upper Missouri; it returns here about the end of July, after completing its trip to the mouth of the Yellowstone and back in six weeks. Your friend could not go and come with that boat; if he goes with it, he must return on a skiff or larger boat of the fur traders, which can always be had. Sometimes a second boat goes up a few weeks after the first (chartered by other companies), but that is uncertain.

Persons well acquainted with that country tell me that the best way is to go by boat earlier (about 1st May) to Fort Laramie; from there to Fort Union on the Missouri, crossing the head of the Mauvaises Terres, a distance of only about 300 miles, and then come down the Missouri River. At Fort Laramie men, horses, pack mules, or carts could be hired, he told me, or he could fit himself out at our frontier and go as far as Fort Lar. with the Emigrants trains. But if he prefers to hire his cart of pack animals at the Fort, which perhaps would be preferable, he could go with some Emigrant train to the Fort or also by stage.

If he prefers to go in June with the Fur Company's (Pierre Chouteau & Co.)[4] boat he must have special recommendations to them, which Prof. Baird in Washington and perhaps also the present Consul General in New York could furnish; letters from the highest authorities of your state might also have their weight. Even if he don't [*sic*] use the boat he ought to have letters to them and from them to their factors at Fort Union. A letter from some military authority to the officer commanding at Fort Kearney and Fort Laramie would also be useful.

He ought by all means go as soon as possible so that he has the summer season before him.

Our Legislature has just passed a law for a geological survey of the state. Who will get it is yet unknown to me. I do not know whether Eastern men calculate on this office. In our state we have only Dr. Prout[5] and Dr. King,[6] who would apply for it. The latter may get it.

Your young friend should be here a few weeks before he starts to make his outfit, etc., and I shall try to assist him as much as is in my power.

Very truly yours,

G. Engelmann

I hope when you come again here your health and the weather will give you better opportunities to see and examine the country.

E.[7]

This information furnished by Engelmann (and possibly derived from other sources as well) evidently assured Hall that a new venture to the Badlands was indeed feasible. By early March he was making rather definite plans, and he wrote to Professor J. W. Bailey[8] at West Point requesting a letter of introduction which Hayden might carry with him for use in visiting Western army posts. Professor Bailey's reply to Hall follows:

West Point, March 7, 1853.

Dear Prof.

Your note of the 2d was duly received but the answer was unavoidably postponed until now. I am glad to hear of you and yours and hope that you are enjoying home after your rambles. I received a note a few days ago from

Mr. Hayden asking about the modes of collecting fresh water Algae which I answered. I have written a note which you will find enclosed, directed in a general manner to "Officers of the Army" which I doubt not will prove sufficient introduction to such Officers as he may meet. I wish you would urge him to look out for anything which may prove "infusorial" in that strange Nebraska or "Mauvaises Terres" region.

I have no acquaintance with any of the Fur Company people, nor have I any influence at Washington with those who might possibly direct that aid should be afforded at the posts which Mr. Hayden may visit. I suspect he will do well enough by trusting to the voluntary kindness and attention of our Officers.

I shall be very happy to receive the 2nd part of your most valuable labours on Palaeontology, which must always be a classical work in American science.

Give to Mrs. Hall assurances of my continued kind regard and believe me very truly

Yr friend

J. W. Bailey[9]

By late March, Hall was considering the possibility of sending a second man to the Badlands in company with Hayden, despite his concerns about the cost. This is mentioned in another letter which he wrote to George Engelmann:

Albany, March 27th, 1853.

Dr. Engelmann,

Dear Sir,

... I have been hesitating about this expedition for some time but finally have decided to carry it forward. I thank you for the trouble you have taken and I shall still feel under greater obligations if you can give Mr. Hayden some advice and assistance after his arrival in St. Louis. He will go up prepared if need be to remain there during the winter. I will give him a letter to you and he will be glad to find on your recommendation some place where he can live economically while waiting & making his preparations in St. Louis. Mr. Hayden is a botanist, or at least has collected this science more than any other. He is ardent, enthusiastic, and untiring in his industry. It is yet uncertain whether I may send some one else with him but if I can send a proper person [I] think I shall do so. I will take steps immediately & have Mr. Hayden fitted out as far as we can do it here & if he needs anything after arriving at St. Louis I will [abide] by your advice & assistance. I hope Dr. Koch[10] may not turn his eyes that way for the present for I want a few of these things for the United States before they all go to Europe, and thus far there has been no systematic work done there as far as I know.

Mr. Hayden has no money and I have none to spare and therefore we shall both be anxious to make the affair as economical as possible.

There are certain points which he can learn in St. Louis I dare say. These are whether the White or Cheyenne are navigable for boats to carry down specimens to the Missouri; and whether it would be desirable to take up a boat for that purpose. Are there not means for returning from there in September so that if two go one could return at that time if necessary? Can you advise me about any thing for presents to conciliate if need be the Indians? Or perhaps this is not necessary.

I am very truly yours,

James Hall

Some of my friends have been advising me to apply for the Missouri Survey but I learn that your Governor will not go out of the state & I don't think the western people like the eastern geologists over much.

P.S. I shall take steps to get letters from the best men here to the agents of the Fur Company and will go to New York to get letters from Cho[u]teau, etc. if there is any one there who has authority. I do not know where to inquire for the principal parties but will learn.

In the mean time I shall be greatly obliged if you can speak to some one connected with the establishment if you think it proper to do so, and perhaps he could suggest some person in this part of the country from whom letters would be desirable, both at St. Louis and at Ft. Pierre.[11]

James Hall also wrote to John Newberry regarding Hayden, apparently seeking further assurance about Hayden's qualifications, since Newberry was Hayden's former mentor. The letters which Hall wrote to Newberry are not preserved, so far as known. But we are more fortunate concerning Newberry's replies, two of which follow:

Cleveland, Mar. 2^{d}, 1853.

Dear Sir:

... In regard to Mr. Hayden I can only say this. In a subordinate capacity Mr. H. would be excellent. He has enthusiasm, industry, and untiring perseverance, but has had little experience in the ways of the world & little practice in the field.

After a little training I think he would serve you as a collector in New York, Ohio, or the Western States very efficiently, and would probably do as well for you in the far West as any one who would take the place of an employee.

I have given him some hints as to the modus operandi of collecting, as well as detailed instruction in the theory. Whether he would satisfy you in the responsible & difficult task of exploring & collecting in the Mauvaises Terres I am entirely unable to say. Only, I know of no one who would probably do better whose services you could obtain...

Yours, etc.

J. S. Newberry[12]

Newberry's second letter to Hall again emphasized Hayden's enthusiasm but questioned his ability to lead the expedition. It also shows Newberry's enthusiasm for the idea, particularly since Congress had not approved David Dale Owen's proposal "for a survey of the far West."

Cleveland, Mar. 23^{d}, 1853.

Dear Sir:

I rec'd yours of Mar. 14th some days since and have only delayed replying to it until Mr. Hayden should return from the country where he has been spending a few days. He has now returned and I enclose a line from him as well as return Dr. Engelmann's letter as you requested.

I am greatly interested in the success of your "Mauvaises Terres" enterprise, and sincerely hope you may be able to carry your plans into execution this season as Congress did not listen to the prayer of Dr. Owen for a survey of the far West and the field is entirely unoccupied. Of course that region will yield a rich harvest to at least the first thorough explorers.

I am sorry your friend has concluded not to go West this year.[13] If of the right stamp his cooperation with Hayden would be of the utmost importance. Hayden would make a very efficient assistant, his industry, enthusiasm & economy would be invaluable in that capacity, but he is from his inexperience less fitted to take the entire enterprise into his own hands than an older person and one who has had more experience.

I should be desirous if I had any design of exploring this region of having a party of several go & that they might remain there a sufficient length of time to obtain something more complete, and a much larger number of specimens than have hitherto been brought from them.

In looking at the maps it seems possible that specimens might be floated down White or Cheyenne rivers, crates (like crockery crates) being used instead of boxes, and the specimens being packed in prairie grass — the flat boat being made at St. Louis or St. Josephs, and taken up by steamboat, thence "poled" up to the Bad Lands. Perhaps all this would be impracticable either from expense or some natural obstacles, but if something of this kind could be done a cargo could be brought, and everything in perfect condition...

Yours in extreme haste,

J. S. Newberry[14]

While thus corresponding with Engelmann, Newberry, and others, Hall wrote also to Hayden, inquiring about the terms under which he would be willing to accept employment. Hayden's reply, offering to work until November, 1853, for $150, follows:

Cleveland, March 3^{d}, 1853.

Dear Sir:

I received your welcome letter yesterday for which I had been waiting some time with considerable anxiety. I was sorry that you could not decide respecting my destiny for the coming season. However, any decision you may think best to make will please me. The Mauvaises Terres expedition seems like a pleasant though arduous one. Still I am ready at any moment to undertake it provided the desired arrangements can be made.

You inquire in Dr. N's letter in regard to my probable success as a collector? I can only say that I shall do what I can. I have had little experience in these matters as yet, and consequently there is a possibility of my not succeeding at all. With all modesty, I think I should exceed even your expectations, for obstacles have no terror to me and I trust never will have....

You wish to know my "terms of agreement." Of course you know something of my circumstances and my wants. I am perfectly poor, in fact a little worse than nothing. I have no resource but myself in time of want and must provide somewhat during summer for winter. I have had the offer of one of the City Schools with a fine salary. But it is not in accordance with my feelings or health, and [I] shall not accept, if I can find employment in the "field" at even one half the sum I could make otherwise. I will make a proposition to you which Dr. N — thinks is a fair one. I will labor for you until the first of November if at that time you can pay me $150, and my expenses until that time. I wish to finish my Medical Studies and if you can think of an equivalent I wish you would inform me. I might do something for others that would be of some value. However you must dictate about that. If you cannot take up my proposals will you please tell me how near you can approach them, or what terms you can make with me, for I cannot give up the idea of laboring for you this summer and even for a term of years if it should [be] desirable on your part. Could I reasonably hope for a situation as instructor in Natural History in an Academy or College I would cheerfully devote all my energies to my favorite pursuits for some years. But as pecuniary matters connected with Science seem uncertain I hardly know what course to take. I suppose that you have not so much money as to be a burden to you, so that it makes me somewhat diffident in stating my terms. Please write me in your next what you think and believe me

Sincerely and truly yours,

F. V. Hayden[15]

Two weeks later Hayden again wrote Hall (for enclosure with Newberry's letter of March 23rd) and reveals his continued suspense, for Hall had not yet given him any final go-ahead:

Cleveland, March 19th, 1853.

Prof. Hall,

Dear Sir,

I wish to send a line to you in Dr. N's letter for I have some anxiety in regard to my destiny for the coming summer. My friends are somewhat anxious to know whether I stay here or go to the extreme West. However, do not hurry matters more than you think best on my account. Of course you will inform me as soon as you decide.

I sincerely hope that our Mauvaises Terres expedition will not prove a failure. It is one which I have most at heart. I have read Dr. Engelmann's letter, also Dr. Owen's Report sent me by Senator Wade, Culbertson's Report, etc., all of which strengthen my confidence that I can perform the tour with success and perhaps with profit. Company would be desirable, but if it cannot be obtained, that would not deter me in the least.

If I go up in June, I might perhaps send down on its [the steamboat's] return a quantity of fossils. In regard to spending the Winter there, it would be rather desirable to me than otherwise, provided I had a proper outfit. I might act in the capacity of physician to some of their companies which I think I could do very well.

I am now somewhat in a state of suspense, not knowing what to do or plan, whether to be making some preparations for a western tour or one nearer home. In case of the former, I suppose I shall need some stout, coarse clothes. I would like some instruments for taking observations, some books, and materials for keeping a full journal and sketching whatever might be of interest. These, I think, can be obtained in various ways without much expense. I would not have these things interfere in the least with the main object of the journey. You will of course advise me as you think best. If I could spend some time there I might make collections in other departments of Natural History, as Fish, Insects, Shells, etc., [for] rare specimens of which Prof. Kirtland and others would give a handsome remuneration. If you thought best, I might perhaps lighten the expense, in that way.

Perhaps you may think I have too many irons in the fire, but an untiring industry will accomplish much.

I have just returned from spending a few days... with Dr. N. The weather was quite bad. We collected some interesting Coal plants, but nothing else important. I think, however, that the weather will soon be sufficiently settled so that I could do a good business at collecting. Have you any work to do in Monroe Co., N.Y.? I have many friends living in and near Rochester, a Mother, Sister, Brother, etc. whom I have not seen but once for 12 years. I should like to spend a little time with them, and if I should, I could labor for you in that vicinity without much expense as I could have some assistance, also a horse and carriage at my command, from my Stepfather who is an extensive farmer in Greece...

I remain truly yours,

F. V. Hayden[16]

The foregoing letter prompted Hall to make a definite commitment to Hayden. Hall's letter is not extant, but Hayden's answer to it tells the story:

> Cleveland, March 30th/53.
>
> Dear Sir:
>
> I have just received your welcome letter and hasten to reply. You need not have hesitated on my account in the least, for I have been ready to go any where you might wish to send me. My ambition will not allow me to fail of success if health is spared me. I am going to Prof. Kirtland's this afternoon, with Dr. N. He will make many preparations for me. If it will not be too expensive, I should like to see you. I think in the end it will pay the expense of going and coming by the benefit I can obtain from your experience.
>
> You will receive this note before Friday and if you think it not best for me to come to you, please telegraph to me, otherwise I shall start from here Friday Eve. and spend the Sabbath in Rochester and be with you Monday Eve. I shall study to make you as little expense as possible.
>
> I have just returned from Prof. K's, and every thing is agreeable. He will pay well for anything I may obtain.
>
> With great respect,
>
> Yours,
>
> Ferdinand V. Hayden[17]

So the die had been cast. James Hall had given the go-ahead to Hayden, and Hayden was only too anxious to accept and get started.

4

Launching an Expedition

Once James Hall had committed himself to engage Hayden, new concerns arose almost immediately. Late in March, Hall received two additional letters from George Engelmann, who was faithfully continuing to pass on information and suggestions from St. Louis. In these letters, Engelmann alerted Hall to the possibility that his projected expedition might encounter competition:

St. Louis, March 19th, 1853.

Prof. Hall

Dear Sir

Some weeks ago I wrote to you about the ways and means to get to the Mauvaises Terres. I learn now that the fur company's boat will leave this season earlier than usually, about the middle or in the latter part of May, and I could perhaps through the intervention of some friends assist your friend in getting conveyance on the boat if it should appear preferable to take that route, and not the land route by Laramie.

I have further ascertained that two German collectors, at present here, propose to go up the same route for the same purpose. I have not been able to see them yet but I learn that they are not naturalists but mere collectors. They intend to go up on the Fur Company's boat.

Dr. Koch… is again in these parts, having settled in Golconda, Illinois, (on the Ohio). I should not be astonished if his sharp instinct would not lead him up there among the Mauvaises Terres, though I do not know that he has any travelling propensities at present.

If I can be of any use to you in this or any other matter, please write to me.

I learn that no appointment has been made yet for state geologist. Dr.

King and Dr. Prout both I learn have chances.

Very sincerely yours,

G. Engelmann[1]

Engelmann's second letter to Hall emphasized the urgency even more:

St. Louis, March 27, 1853.

My Dear Sir

About 4 weeks ago I addressed you a letter in answer to yours. I have since learned that the fur company's boat will leave this season earlier than usually, so that your friend must look out sooner if he intends to avail himself of this opportunity.

But I write principally to apprise you of the fact that there are here at present two German collectors, not naturalists, but assiduous collectors, who having heard of those riches, propose to go to the Mauvaises Terres and will probably obtain permission to use the fur company's boat. Whether that will have any influence on the movements of your friend, and what influence, I cannot tell. At all events he should form his resolution soon, and be here in good season.

It is said that Prof. Swallow, Prof. of Chemistry at the State University at Columbia, in this state, has obtained the appointment as state geologist.[2] He is said to be a good chemist, but a few months ago did not pretend to geology at all!

Yours in haste,

G. Engelmann[3]

In Hall's reply to the second of Engelmann's letters he again mentioned his inclination to provide Hayden with a companion, but hesitated on this point because of the added expense:

Albany, Ap. 10th, 1853.

Dr. Engelmann,

My Dear Sir,

I thank you for your favor of the 27th which I would have answered sooner but have been waiting a few days for Mr. Hayden, the young man who proposes to go to the Mauvaises Terres. He will be in St. Louis as early as the first of May and will have his preparations made as far as he can make them. I will rely upon you for advice after he arrives there.

Mr. Hayden is an ardent collector and has pursued botany to some extent, but has no knowledge of a wild country. He goes out at my expense and I am desirous to avoid unnecessary expense while at the same time I wish to give him all possible facilities for making collections.

I had supposed that at Ft. Pierre he could find one or more men who are acquainted with this mode of life, and know something of the Indians and can speak their language. It might perhaps be wise to take a man of this character from St. Louis if such an one could be found there. Still I must rely upon your advice and Mr. Hayden will act accordingly. I have a strong inclination to send a second person from here, but I do not know how expensive it will be to do so.

I am going to see Mr. Cho[u]teau who is now in N. York and get his consent and approbation for Mr. Hayden's passage etc. on the boat.

If the boat is likely to leave before the middle of May please write me or telegraph me at once that I may apprise Mr. Hayden who will be at that time in Cleveland.

In St. Louis he will procure his provisions and any outfit which he may not have already obtained.

I shall feel obliged if you can refer Mr. Hayden to some boarding house or Hotel where he may live without too much expense while in St. Louis.

If you know any person familiar with the Mauvaises Terres and the preparations necessary for such an outfit it would greatly facilitate Mr. Hayden's objects to become acquainted with him.

For any aid or advice you can give him in this matter I shall feel greatly indebted, and shall hold myself in readiness to reciprocate the favor.

Can you tell me if these German collectors are such men that Mr. Hayden could with propriety make any arrangement to travel with them? How long do they intend to remain up there?

With many thanks,

I am very sincerely yours, etc.

James Hall

P.S. I am surprised at the selection of Geologist by your Government. This gentleman is quite unknown to any one here as a geologist.

I presume on many accounts it would be better to procure men at Ft. Pierre if that can be done, for if Mr. Hayden remains through the season there is no way for men to get down in the autumn.[4]

On April 10th Hall also wrote to Dr. H. A. Prout, of St. Louis, for such information as he could give about the costs of travel to the Badlands, and he also inquired about Dr. Prout's availability for participation in the expedition. The doctor replied:

April 27, 1853.

Prof. Hall

Dear Sir

I rec'd your favor of the 10th Inst and would have answered your enquiries could I have seen Mr. Sarpy before this time. Mr. Sarpy is at the head of the American Fur Company, being a principal partner. I learn from him that the following estimate of the expenses of a trip may be regarded as very nearly correct. Passage of one person to Fort Pierre or Mauvaises Terres fifty dollars. The employment of two men for two months at 2$ per day — 240 Dollars. The purchase of 4 mules, one for carrying specimens 400 Dollars. Return and transportation on specimens 250 Dollars. He thinks that 1000 Dollars may be set down as a fair estimate allowing for all contingencies. Men cannot be had at these posts for less than 2$ per day for a short time but if employed by the season at from 1 to 1.25. They could be employed here at thirty dollars pr month but their passage would make it nearly the same as giving 2 Dollars per day. The mules could probably be sold for nearly what they cost so that the expenses would be reduced to 650 Dollars or 700.

Nothing would give me greater pleasure than to go on such an excursion but my limited means would not justify it as I am compelled to attend to my practice for a support. My practice is lucrative and I could not give it up without much sacrifice of pecuniary interest so that I must content myself to look on and see this good work carried on by others. Mr. Sarpy has assured me that he will give every facility to your agent if you should send him with his expedition which will leave here about the middle of May. He has promised me moreover to see the rights of American geologists secured against the foreign collectors who have or will be sent out this season. We ought to explore this region first and then leave them to glean as much as they please for foreign cabinets.

I remain Sincerely

Yours

H. A. Prout[5]

Hayden's proposal that he and Hall meet at Albany to discuss their plans met with Hall's approval. En route to Albany, Hayden visited his family and friends at Rochester, and he stopped there also on his return trip. The conference of Hall and Hayden at Albany was, from the available information, cordial and profitable. Back at Cleveland, while beginning his preparations for the journey, Hayden received the following letter from Hall:

Albany, Ap. 19, 1853.

Mr. Hayden

Dear Sir

. . . I sent your package to Rochester, enclosing one or two small articles beside the india rubber cloth. The chart I thought might be useful sometimes, if you find a person wishing to know something of geology.

I enclose a list of a few things, all of which I think you have on your memorandum. The wrapping paper is important and do not get straw paper.

Dr. Newberry knows what you need. I enclose you a check for $50 and I think you need not take any more money with you.

In case you want more you can write me but I think the better way will be to draw on me, which you can do at Ft. Pierre, through Cho[u]teau & Co., giving me notice by letter if you can. Mr. Cho[u]teau and his partners will do all they can for you. Mr. Culbertson — the brother of the one who wrote the journal of a tour to the Mauvaises Terres — will probably be on the boat with you.

Mr. Cho[u]teau thinks the boat will not leave sooner than the 20th of May and that ten days will be ample time to make all your preparations there. I presume this will be so. He has written his friends in St. Louis that you are going up and if you arrive there before him they will give you all necessary attention…

I think you will do well to find Dr. Engelmann even before removing your luggage from the boat, as I think he will have some place of stopping to recommend to you. His office will be readily found. He is one or two streets back of the Planters house and about one block farther south…

I have also written Dr. Litton[6] to inquire of Mr. Evans in relation to his journey. I wish you to write me as soon as possible after you get in St. Louis. Let me know the prospects of your journey, etc.

Mr. Cho[u]teau thought it might be better to take a man with you from St. Louis, while you could get mules & a guide at the fort. Perhaps also, he says, there may be men enough to be spared at the fort for all you would want. If so that will be the better course. All this you will not fail to learn in St. Louis. If there be any difficulty, I could send a man from here.

Mr. Cho[u]teau says you can leave your collections there safely over winter to come down by the boat in the spring. He thinks you will find the winter a long one if you remain. Still if you come down it will be necessary to buy a boat and have men. Again if the Germans remain you might if agreeable all come down together and make less expense. It may even be worth while to consider whether you would not come down in the fall & return again in the spring. If you came down you could bring a part of your collections and leave the others etc.

Sincerely yours

James Hall

P.S. Have a letter from Dr. Leidy. He is anxious about the Mauvaises Terres since there are two German collectors going up — perhaps may send some one to go with you.

Will enclose a letter of introduction to Mr. Cho[u]teau in a day or two.

A list of supplies and equipment was enclosed with James Hall's April 19, 1853 letter to Hayden:

Wrapping paper — good and strong — two or more sizes. 20 reams or more.

Glue.

Pocket level if possible.

Tape line.

Spade.

Pickaxe of the form of a common pick axe but lighter and the edge made three inches wide like a mattock — and the other end having an edge of equal width or less turned at right angles.

[Sketch, with notations]

This is an important tool.[7]

The next day Hall wrote letters of introduction to Pierre Chouteau, Jr. and George Engelmann which read as follows:

Albany, Ap. 20th, 1853.

P. Cho[u]teau Jr. Esqr.

Sir

I take the liberty of introducing to you the bearer, Mr. Hayden, the young man of whom I have already spoken. He is now on his way to the Mauvaises Terres and in many things must necessarily be dependant upon facilities, to be obtained only through the favor of yourself and friends. I beg to assure you that his object is entirely a scientific one and in that respect I trust it will commend itself to your favor.

I am very respectfully

Your obedient Servant

James Hall[8]

Albany, April 20th, 1853.

Dr. Engelmann

Dear Sir,

I beg leave to introduce to you Mr. Hayden, the bearer of this letter of whom I have already written you. He will be desirous of your aid and advice in preparing his outfit for the Mauvaises Terres.

In reference to taking men with him he will be governed by your advice and the information which Mr. Cho[u]teau may have about men at the Fort. Mr. C. told me it was not improbable that men enough could be obtained at the fort.

Any attention you may be able to show Mr. Hayden in this matter I shall regard as a personal favor for which I shall feel under many obligations.

I am very sincerely yours,

James Hall[9]

About the same time, F. B. Meek, Hall's assistant, left alone at Albany while Hall was away on a trip, penned the following undated letter to Hayden at Cleveland. It was probably written about April 23rd. When he wrote it, Meek evidently had no assurance — or even hope — that he too would be sent to the Badlands:

> Mr. Hayden
>
> Dr. Sir;
>
> Prof. Hall has just started to Pennsylvania, and requested me to say to you that you should provide yourself, before leaving St. Louis, with a ball of twine of the size of the enclosed sample. You will need it in fixing your tent.
>
> Dr. Leidy has written Prof. H. a long letter containing some useful hints in regard to collecting fossil bones, which he will communicate to you before you start. Dr. L. is much annoyed at the idea of the German collectors visiting the Mauvaises Terres, and speaks of sending a man with you, in order that you may be able to send back to the states a good collection before they return.
>
> With my best wishes for the success of your expedition, and hoping that you may find all manner of interesting things in the far west,
>
> I remain sincerely yours,
>
> F. B. Meek
>
> P.S. Don't forget to write.[10]

In this note, Meek alluded to the following letter which Hall had just received from Leidy, and the substance of which Hall later communicated to Hayden:

> Philadelphia, April 20, 1853.
>
> Mr. Hall,
>
> My dear Sir,
>
> I am glad to hear you are going to send Mr. Hayden to the Mauvaises Terres, which with its remains I view as a darling child. The information of two German collectors going there this summer has annoyed me exceedingly, and I am delighted to hear you possess the same patriotic feelings as myself. I had fully intended going myself this Spring, but the death of Dr. Horner[11] having made a vacancy in the University I was much disposed to remain at home and try my chance to the succession. I think I could raise a few hundred dollars at the Academy towards sending another person, and if you write to Prof. Henry,[12] I think probably the Smithsonian might contribute some aid. I also feel satisfied that a collection of duplicates would bring a good price.
>
> In regard to advice to Mr. Hayden, I respectfully suggest the following:
>
> 1. The most important specimens are <u>Crania</u>, <u>teeth</u>, and their <u>fragments</u>.

2. If the specimens are enveloped, more or less, in a matrix let them remain so, for they carry without injury.

3. Bring no fragments of large bones.

4. Bring no fragments of the long bones whatever, except the joints or articular ends of those which are small in size.

5. Bring no fragments of turtles nor the largest ones even if nearly perfect.

Of course the above advice is intended for a small party. The specimens are heavy and a few horses would soon become loaded even with those most characteristic. I have here 500 lbs of entirely uncharacteristic and worthless fragments which were brought by different collectors from Nebraska, and yet they say some fine specimens were left behind because they were already loaded.

What are the facilities of going out there and returning in the autumn? A gentleman in this city would go at his own expense if he could return in September.

My Ancient Fauna of Nebraska is now going through the press, and the plates, 24 in number are all finished. The plates are much better than those in the Geol. Rep. of Dr. Owen, as you may judge by the specimens sent.

With great respect

I remain at your service,

Joseph Leidy[13]

Following this urging by Leidy, and the possibility of some financial help, Hall finally concluded, doubtless with great reluctance, to send his assistant and right-hand man, F. B. Meek, along on the expedition with Hayden. This decision, and his reasons for arriving at it, are expressed in a letter to Engelmann:

Ceres, Allegheny Co., N. Y., Ap. 23[d], 1853.

Dr. Engelman,

Dear Sir,

I wrote you a few days since, but since that time I have concluded to send Mr. Meek, my assistant, with Mr. Hayden, up the Missouri. I am led to this partly from the fact that Mr. Hayden has had no experience of this kind of life and partly from the consideration that it is important now to secure a good collection of these things or they will soon all be gone to Europe. Dr. Leidy has written me strongly on that point, and thinks a good collection might be sold to pay a part of the expense. At any rate I am committed now and the expedition must go properly prepared. I have also a mind to send along a young man as a laborer to assist. Such man will be needed and must be hired either in St. Louis or at the forts. Mr. Cho[u]teau thought it possible that such could not be had at the Fort but could no[t] certainly inform me, saying they would know in St. Louis. I presume, however, it may not be more expensive to send one from here than to hire one in St. Louis or at the fort. If you have any information on this point please communicate to me at Albany where I shall return in a week.

I can hire a man for about $12 per month paying all his expenses, and of course I should pay the expenses of a man after hiring him at St. Louis or elsewhere. Perhaps you can advise me.

Mr. Meek will probably arrive in St. Louis about the 15th as Mr. Cho[u]teau thinks there is no probability [of] the boat leaving before the 20th May. Mr. Cho[u]teau will return about the first of May and if any change of time is made I beg you to telegraph me. The boat will not leave before he arrives certainly.

I commend Mr. Meek & Mr. Hayden to your attentions and remain

Very sincerely your much obliged & obdt servant,

James Hall[14]

Meanwhile Hayden was at Cleveland and busy with his preparations, including the purchase of supplies. In this he was aided by Newberry. Thus engaged, he suddenly received word from Hall to the effect that Meek, too, would go on the expedition:

Ceres, April 24, 1853.

Mr. Hayden

My dear Sir —

I wrote a line to Dr. Newberry last night after arriving in Olean to ask you to wait for a letter from me. I had written and enclosed letters, from Albany, for you, to Cleveland. I have heard from Dr. Leidy who feels very much annoyed at the idea of the two German collectors going there and is as anxious as I am that a good collection shall be secured for our own country. Under all the circumstances I have concluded that it is better for Mr. Meek to accompany you. He has had experience in the kind of life and that will be of the first importance. His acquaintance with Indian character and customs is also important. Mr. Meek however will wish to remain at Albany as long as possible and you will perhaps do well to prepare yourself and leave for St. Louis to be there by the 10th. The boat leaving on the 20th according to Mr. Cho[u]teau, this will give you time to get prepared for the expedition. Mr. Meek will leave Albany about the 7th or 8th and arrive by the 15th or sooner and have five days. Even this time will be enough for all preparations. Still it may be well for you to have a few days to see some persons and when Mr. Meek arrives, you can together make your purchases of provisions, etc. for the voyage.

As to the matter of remaining there during the winter it can depend upon the success of the expedition, and the facility of getting down. This must in a great measure depend on what you may learn at St. Louis and at the fort. The result must be the finest collection ever obtained and now that I am fairly engaged in it I shall take all necessary steps to have it succeed. I regret to part with Mr. Meek for six months for I need his assistance here. Still I think it important that you have some one with you.

I shall enclose letters to you care of Dr. Engelman, St. Louis.

If there should be anything in the way of preparation which you have not done when you leave Cleveland I will have it ready and send by Mr. Meek. He will have little baggage and can take a few extra things if necessary. Should you decide, before leaving St. Louis to return this autumn, you had better go as light as possible, though the weight of your books will not add much, I suppose... If you have not all the tools you want, I can get them in Albany when I return and send by Mr. Meek.

Should you open your box at Cleveland you will find in one pair of your boots a pair of James' slippers which will be useless to you and you can leave them with Dr. Newberry.

You can go to St. Louis by two routes, one by Chicago to Peru on the Illinois River and thence by boat to St. Louis; the other by Cincinnati and thence by boat. You will probably find one less expensive than the other. Of this you can learn in Cleveland.

You will prepare in Cleveland all that you can in the apparatus for your scientific objects and after Mr. Meek arrives in St. Louis there will be time to get provisions, etc.

Mr. Dana says he would like to get any crawfish or crustaceans of any kind you may meet with in your travels.[15] You will give your first attention to the crania, teeth, etc., and leave other objects as of secondary importance. Should you conclude to remain through the winter, you will have abundance of time the next season as I doubt not you could then come down with Dr. Owen or some other parties who will go up to come away in October. This plan is worth considering as you could then get to Albany in time to attend the course of lectures there for the winter and there is also now a second course in the spring. A plan of this kind would give you an opportunity of taking a greater range of country and of embracing more subjects. Were I as free as you are I should be inclined to do so unless the expense of remaining over winter is too great, and that it would be economy to go out again in the spring.

You will find it much pleasanter to have Mr. Meek for a companion and I shall now feel less solicitous than if you had started entirely alone.

Give my kind regard to Dr. Newberry, and believe me

Very sincerely yours

James Hall

Should you receive this in time, direct a letter to me at Olean. I shall leave there next Saturday or Monday and Buffalo a day or two later. I will inquire also in Buffalo.[16]

Hayden wrote Hall three letters while in Cleveland to keep him informed of his progress. In these letters he stressed the economy he was practicing in preparing for his journey:

Cleveland, April 26th/53.

Dear Sir,

I arrived in C. last evening safely, but with some hindrances. I started from Rochester Saturday Eve and on arriving at Buffalo, I thought for the sake of a good night's rest I would go to C. in a boat as one was ready to start in half an hour. A storm arose in the night so that the boat could not go into port and so she went on to Toledo. Thus I was hindered 2 days.

I have been laboring very hard today purchasing numerous articles, with good success. I think I shall be ready to start from here Thursday Eve. I am somewhat anxious to get to St. Louis for several reasons, and I think my expenses will not be much greater there than here...

I received your check for fifty dollars, and received the money today. Dr. N. thinks I had better get the more bulky articles at St. Louis so as not to be annoyed with too much baggage. I find it somewhat expensive. I shall probably get a part of my wrapping paper here and a part at St. L. The difference in price will hardly pay transportation. Good paper is $1.75 per ream. I have kept an account of all my expenses and purchases so that you can see it on my return or before if you wish.

The expedition, it seems to me, is going to be rather expensive, but I will make the best of it possible. The more I think of it the more enthusiastic I feel. As to aid you spoke of, I think I can get along alone until I get to the fort, though as you think best. As to staying through the winter it seems to me now that I should prefer it, though I can tell better at St. Louis, I suppose.

With the advantages I shall have from the Fur Company, etc., you may be certain that the German Collectors will not get much the start.

I shall if possible send down as great a variety as possible.

I have received Mr. Meek's note. Will write to him from St. Louis. I have about given up the idea of taking a thermometer, etc., and I do not know as it will pay expense. I certainly ought to have a good gun. Perhaps some one will furnish me one on my way. Perhaps I am starting for St. Louis sooner than necessary, but I am anxious for fear of some disappointment. Please remember me kindly to your family, and I remain with great respect,

Yours,

F. V. Hayden

Hayden's second letter to Hall from Cleveland, written two days later, updated Hall on his progress:

Cleveland, April 28th/53.

My Dear Sir,

I have just received your letter from Olean and reply immediately. I have intended to start this evening for Cincinnati, having made all preparations that are thought necessary here. I have as yet got no wrapping paper. Dr. N. thought it might be obtained as well at St. Louis and the difference in price would not pay the expense of transportation. However, you can judge about that and if you prefer to get it here, Mr. Meek can bring it. Perhaps the package you spoke of had better be made in A. and any other things of that kind that may be necessary.

I have already written to A. to you. As far as Botany is concerned, I think Dr. Engelmann will assist me better than anyone else. I am anxious to get to St. Louis so as to meet him and get all the information possible and of any others that would have information to communicate. However, I may not start until tomorrow eve or next week.

I shall be very glad to have company, and the reason why I did not say it was I supposed that you did not wish to make the additional expense… Mr. Meek will be a very agreeable companion, an exceedingly modest unassuming man, one with whom I should feel willing to share any discoveries that may be made.

As to staying through the winter I would say, There is but one question in my mind, Will it pay expense and advance the success of the expedition? If so, no regard for personal comfort or anything of the kind would deter me from so doing. However, either way will please me. I am willing to stay as long or as short a time as you may see fit and find profitable to you. It would please me to make a thorough exploration of that region if it could be done without too much expense. My great anxiety now is to make up as much as possible my want of experience by getting the experience of others.

Truly yours,

F. V. Hayden

Hayden wrote Hall the third letter the next day, just before he left Cleveland for St. Louis:

Cleveland, April 29th/53.

Dear Sir,

Inclosed are specimens of the wrapping paper and the amount I take with me. 20 reams would be so bulky that Dr. N. thought I had better wait and see what could be done in St. Louis. My baggage has already accumulated so that I fear it will be somewhat expensive. If anything more is needed from Cleveland Dr. N. says he will forward it to St. Louis. I shall start tonight at half past 8 for Cincinnati. I feel anxious to be on the ground, and I think my expenses there will not be much greater than here. If I have any leisure hours I shall spend them in making collections in that vicinity. Dr. N. has been very kind in assisting me and I can think of nothing more to get here.

The route of Cincinnati is the best as far as I can learn. If I go by Chicago I shall have to go by Stage 15 or 20 miles which will be inconvenient with my baggage. I cannot learn how long I shall be going from C[leveland] to St. Louis. But I suppose 5 or 6 days.

I mailed a letter to you at Olean yesterday. Have received all your letters of introduction. I shall write to you the day I arrive at St. Louis and then I can tell better the difference in the price of articles here and there.

With great respect,

Yours,

F. V. Hayden[17]

Hayden did not actually wait to write to Meek until he reached St. Louis, but dispatched this note to him from Cleveland:

Cleveland, April 29th/53.

Friend Meek,

I received a letter from Prof. Hall yesterday morning stating his determination to send you with me. How does that suit you? It suits me right well. It seems to me that two would do more than twice as much as one alone, and the great object of the expedition is to obtain as great a collection of fossils as possible. As Prof. Hall says, "a far greater collection than has been made before"...

Those tools that Prof. Hall mentions in his letter I shall have to get made in St. Louis if they are not made in Albany.

When you come... go by way of Buffalo. Don't take a boat but go on the cars and then you will be more sure... I was 2 days and one night longer coming than I should have done. That will make some difference with you if you start late. To avoid such delays I wish to have time enough.

I shall take down a large box of very fine wrapping paper, not more than half enough I suppose. If Prof. Hall prefers to get [it] before you go I suppose you will bring it, or it can be got at St. Louis. I have 34 ½ pounds of very fine stout paper 30 x 34, and 4 reams of smaller size and more pliable. I shall have as much baggage as I can get along with. It will amount to some 400 or 500 pounds weight. If I should need anything more from Cleveland, Dr. Newberry will send it to me.

Wishing you a good journey from A. to St. Louis I remain

As ever yours,

F. V. Hayden

If you bring along your paint brushes, we may make drawing of shells on rainy days and give me a chance to learn something of the art.

H.[18]

So Hayden was on his way, with his friend Meek soon to follow. Hayden reached St. Louis on May 9th, and reported his arrival to Hall by telegraph. Hall acknowledged this message with a note in which he informed Hayden that he was giving the direction of the expedition to Meek. Hayden did not receive this letter until a week later, however:

Albany, May 9th, 1853.

Mr. Hayden,

Dear Sir,

I have just received your telegraphic d[i]spatch. Mr. Meek left this morning and I hope will be on the ground before the boat leaves.

I will mention now one point that may be of interest to Mr. Meek and yourself. It may sometimes be necessary that the direction of your expedition and the affairs connected with it be under the direction of one of the party. I presume that you can have no objection to Mr. Meek exercising this prerogative from his greater age and his experience in the field. I presume you will have no difficulty however in arranging this matter between yourselves. Still I think were I in your place I should tender the direction of all matters to Mr. Meek as soon as he joins you. I think you both together can work to far more advantage than one alone and with strangers only.

I hope you may have a pleasant and successful journey.

Sincerely your friend,

James Hall[19]

At St. Louis, while awaiting the arrival of Meek, Hayden sought out Engelmann, Evans, and others who could give helpful information about travel routes, equipment, provisions, and other matters pertaining to the expedition. In making these rounds, Hayden gained his first acquaintance with St. Louis, then the gateway to the Far West, and an introduction to men prominent in western affairs with whom he would have important connections in the future. All the while, he was making preparations for the expedition. From a long letter he wrote to Hall reporting his activities, it is clear that Hayden was on excellent terms with Evans:

St. Louis, May 10th, 1853.

Prof. Hall,

Dear Sir,

Dr. Evans has just arrived, on his way to Oregon. He will first go to the "Mauvaises Terres," make a collection, send them down the River, and then proceed on his expedition. I spent most of last evening with him. Father DeSmet has gone[20]... Rev. Murphy will be absent 2 months. I cannot see either of them.

I have been very busy getting information as to the best course to pursue. From the best information I can obtain I am satisfied that the land route is far preferable. That is, go by Steamboat to St. Josephs, fit out there and go to Fort Laramie and thence to the "Bad Lands."

I spent some time with a Mr. Campbell... who has spent many years in those regions, and owns a trading post high up above the Yellow Stone and [is] a great trader.[21] He very strongly recommends to use the land route as far cheaper, more pleasant, and more certain of success. He has promised to give me letters to different men of his organization who will assist me on my way.

But I can make no decision until Mr. Meek arrives and [I] hear something from you. I have not as yet heard a word and it is a source of a good deal of perplexity to me.

I will explain the reason why I prefer the land route as far as I understand it now. If we go up in the boat to Fort Pierre, we cannot make the collection and return in three months for less than a thousand dollars, Mr. Sarpey [*sic*], one of the partners told me.[22] If we do the land route the same amount of money will keep Mr. Meek and myself 6 months, bring him back, and continue my labors a year longer. If this is true (and it appears so to me) it is a matter worth considering. If we go the land route, we go from here to St. Josephs by steamboat carrying our provisions, baggage, with a good light Conestoga wagon. At St. Josephs we get two good mules, put our effects into our wagon and go with the numerous emigrant trains which constantly fill the road. After leaving St. Josephs our expenses will be nothing. We travel through a fine country in which a thousand things of interest may be found, and as safely as on the boat. From St. Josephs to "Bad Lands" 750 miles, at 25 miles per day, 30 days. Mr. Evans thinks we can get there as soon or sooner than he will from Fort Pierre. When we arrive there we have independently of anyone all our fixings, two good mules and a good wagon. At Fort Laramie we can get plenty of help cheaper and better than at Fort Pierre. Many of the soldiers or even officers will like to relieve the tediousness of situation in that way. If not, there are a plenty of voyageurs that may be obtained. At Fort Laramie, I am told, we can live much cheaper, and in all probability [it will] cost me nothing to stay through the winter. Our "treasures" we can transport to Fort Pierre and have them sent down by boat or they can be sent from Fort Laramie in the Quartermaster's wagons which always return empty. Mr. Evans thinks that will be the better way and that he will probably send his collection down in that way.

If we go up by boat, the company will furnish our supplies, charge us 50 dollars each for our passage up [and] charge us for our freight. When we get there they will furnish us men at from 2 to 3 dollars per day and an old cart that will break down the first miles, at one dollar per day, 2 mules that have made out to pinch through a cold winter with a small remnant of life left for one dollar per day each. They will expect to furnish us food for ourselves and team, at about three times its real value and poor at that. Every thing we get, we must pay at least three prices for. They are willing to afford us all the facilities in their power. But how do they do it? Why, for every thing that they do or furnish, we must pay them an enormous price, and make as little trouble as possible.

It is certain that the Germans are going (i.e. to the Bad Lands) on the boat. They have made all their arrangements, paid their passage, and secured the company's assistance. This they did some months ago. Mr. Evans did the same, though he says that if he had not been delayed so long in Washington he should have taken the Land Route. Now we come in for our aid after them. They select their mules, men & carts and then we will stand our chance.

As to our being behind hand, there is a chance for us, if we are diligent, to get there even before they do, and if we should not, Mr. Evans says that after we have all labored two years those that come after us will have just as good a chance and perhaps better even. But a small portion of the Bad Lands has been explored, and, he says, there is no reason why there should not be other localities as good yet to be found.

Mr. Evans goes to the Mauvaises Terres by direction of Prof. Baird, which may account for his not noticing your requests.

I wish you would consider what I have written and if you think best, telegraph to Dr. Engelmann. Otherwise when Mr. Meek arrives we will do what seems best. I prefer the land route with all the risks, and I prefer to go independent if I have to sit up all night and watch the mules and sleep in the wagon days, rather than let the Fur Company speculate out of me.

I have been busy getting instructions and making arrangements as far as I could. It is much to be regretted that Mr. Meek could not have started at least a week earlier. However I will have matters so that a day or two will fit us out to go either route as it may seem best. If we go by land there must be some different arrangement in regard to funds. It will cost about 350 dollars to fit out and get started from St. Josephs. After that, the expenses will be trifling, until we arrive at Fort Laramie. After we get through with our team and things, they can be sold for at least one half their first cost. Dr. Engelmann has just inquired the cost of a good wagon. He finds a man who will make a good wagon, fit it all out with cover, etc., warrant it, for 70 dollars. He needs but a week's notice to have it ready... We need a letter from the Secretary of War to the Quartermaster like the one in that little Book of Instructions of Baird's, and sent on immediately. Dr. Engelmann will see that it is sent to Fort Laramie.

I have written as I think at present. Perhaps when Mr. Meek comes, things will look different. I shall be satisfied either way. Dr. Engelmann wishes to be remembered to you and will write when I leave.

I have not written very plainly for my pen is not as good as a stick. I shall write again soon.

Truly yours,

F. V. Hayden[23]

Meek left Albany on May 9th for St. Louis, and he carried with him the following letter of introduction:

Albany, May 7th, 1853.

P. Cho[u]teau, Jr., Esqr.,

Dear Sir,

I have taken the liberty of introducing to you the bearer of this, Mr. F. B. Meek, who goes out to join Mr. Hayden in the scientific expedition to the Mauvaises Terres. Mr. Meek has already had experience in this kind of life having been with Dr. Owen on his late surveys of the Northwest, and on this account as well as to give force to the expedition I have concluded to send him.

May I ask the favor of your kind offices in giving to Mr. Meek and Mr. Hayden such facilities as are consistent with your own arrangements and means for pursuing their journey to the Mauvaises Terres.

I am with great respect,

Your obdt servant,

James Hall[24]

Meek reached St. Louis five days later, at about midnight on Saturday, May 14th. Shortly after his arrival he received a letter from Hall, who was uneasy because of Dr. Evans' intention of returning to the Badlands, a subject about which Hall had been making inquiries:

New York, May 11th, 1853.

Mr. Meek,

My dear Sir,

...Mr. Cho[u]teau says he has written to the firm in St. Louis to give you all the facilities in their power. He thinks that without doubt there are men at the fort to be had and men who are acquainted with the country. He mentioned two who have travelled over this region extensively. I think therefore you will have no difficulty in arranging all matters of this kind after your arrival at Ft. Pierre.

I shall write to Major Stevens[25] and explain matters and I hope there may be no interference between Dr. Evans' objects and your own.

It is scarcely necessary to say anything further, and perhaps it may all be well enough yet, though I cannot quite understand the object of this long silence on the part of Baird and Evans.

It is possible you may have time to see Dr. Litton and if so I wish you would make some inquiries about the Missouri Survey. If my terms are agreeable to them we shall have the fossils to study, and I presume this will be a matter of as much interest to yourself as to me. If you can give the object any aid through Dr. Litton I trust you will do so.

In haste,

Sincerely yours t'y,

James Hall

Baird writes me to recommend an assistant for one of the Pacific Railroad Surveys for a year or a little longer. I presume it may be similar to the one for which Evans telegraphed you. I wrote him that you were engaged with me and I had no knowledge or experience of any one who would be a proper person.[26]

Hayden wrote Hall once more, on May 16th, reiterating his willingness to leave the direction of the expedition to Meek. From this time on, however, Meek assumed (along with the other responsibilities of leadership) the task of keeping Hall informed of their progress. Here Hayden mentions all the competitors who are in St. Louis and are doing their best to derail the expedition that his heart is set on:

St. Louis, May 16th, 1853.

Prof. Hall,

Dear Sir:

I received your letter dated May 9th, this morning. It was the first word I have had from you since leaving Cleveland. Mr. Meek arrived here 1 o'clock Sunday morning. We have been making our purchases and shall be ready for the boat which will not start until the eighteenth. I had already adopted your suggestion of giving the whole charge of things to Mr. Meek. He has had experience in such matters and I have had none. He will be more cool and calm, and will make his past experience available. I had made up my mind from all the information I could receive, that the Land Route by Fort Laramie would be the best, and I know that it would not cost half as much, still there are reasons which Mr. Meek has for going up on the boat which he will tell you. The expense will be very great, 1000 or 1200 dollars this fall. I fear the collection will not equal the expense by far. Mr. Meek has refused here the most flattering offers to leave you and join Evans & Major Stevens. He has certainly acted nobly, truly so, showed himself to be a man of the sternest integrity. He deserves much credit, and should have the credit of this expedition…

Drs. Norwood,[27] Evans and Shumard[28] are here. Shumard & Evans go to Bad Lands. Dr. Owen will be here soon. Major Stevens and his party are here. All are most violently opposed to our going to Bad Lands. Mr. Meek will tell you more fully. Prof. Swallow is here also…

F. V. Hayden[29]

Thus Hall's letter of May 9th specifying that Meek should lead the expedition had finally reached Hayden, and Hayden was very gracious in accepting his subordination to Meek's leadership.

5

Underway at Last

One might suppose that the partners, Meek and Hayden, now that they were reunited at St. Louis, could complete their preparations in peace and, on the eve of their departure, relax a little to enjoy the prospects of their long awaited adventure. Not so, for others were now in St. Louis eager for their own expedition. And as hinted in the previous letter, serious difficulties developed, which no one seemed able to resolve to everyone's complete satisfaction. Referring to these troubles in his biography of James Hall, John N. Clarke wrote:

> Doctor Evans was also in St. Louis, outfitting for another expedition into the country for a Pacific Railroad Survey into the Northwest, cooperating with Governor Stevens of Missouri and the geologist, Benjamin F. Shumard. The young men [Meek and Hayden] found they were not welcome companions to this party on the only boat that was going up the river that season. The Government party wanted no competition and Governor Stevens was quite forbidding in his attitude toward the new comers. Dr. Engelmann, high-minded and large-hearted, espoused their cause and by lucky chance [Louis] Agassiz[1] happened just then to be in St. Louis lecturing to the Academy of Science and by his great influence helped to satisfy the Evans party there were fossils enough for all. Hall himself, with characteristic procrastination when payment of money was involved, nearly wrecked his own plans by failing to send on a remittance for expenses...[2]

So Meek learned that the distinction of leadership could carry with it sore tribulations. Just before the steamboat started up the Missouri River, he wrote Hall a detailed account of what had transpired:

St. Louis, May 19, 1853.

I now sit down after a world of perplexity to write you a hasty line.

I did not write sooner because I wished to wait until I had settled, in my own mind, the propriety of going on with the expedition, under the circumstances in which I found myself placed. On arriving here, I found Drs. Evans and Shumard were going to stop on their way to Oregon to make a collection of fossils at the Mauvaises Terres. I called upon Major Stevens and found him very much offended at you, for sending out the expedition, but upon being assured by Mr. Hayden and myself that you knew nothing about Evans' intention to visit that region this season, he appeared to view the matter in a different light. I have since read to him a portion of your letter of the 9th inst. and he said he was satisfied and took back all he had said about you.

Being at a loss how to proceed I telegraphed you, and waited some time for an answer, but have received none. After thinking the matter over I concluded the best thing I could do would be to submit the whole matter to Prof. Agassiz and Dr. Engelman[n], and be governed by their advice, for I am well satisfied they are your sincere friends. They consulted about it for some time and had a conference with Dr. Evans, after which they advised me to abandon the expedition, on the conditions that you should be compensated for the amount you have expended and that Mr. Hayden and myself should be taken into their corps as assistants. (They were anxious to engage me.) Prof. Agassiz said their reasons for this advice were that they, as your friends, considered it of more importance to you that the whole matter should be amicably settled, and your money paid back to you, than the value of my services for the coming year would be to you in New York. I showed Prof. Agassiz the obligation between you and myself, and told him I felt a great deal of delicacy about such a step. He answered that the relations between you and himself were such that he would not hesitate to take it upon himself to act for you in your absence, in a matter of this kind, and that he would see you on his way to the east and explain the whole matter to you so that you would be satisfied. Accordingly this proposition was submitted to Maj. Stevens and Dr. Evans, and after some consultation they answered that they declined it as far as Mr. Hayden was concerned, but that they would agree to all the other terms. Upon some little reflection I declined their proposition. Early on the following morning a different proposition was made [to] Mr. Hayden and myself which we promptly declined and it was immediately withdrawn. This left the whole matter exactly where we started. After waiting another day to see Prof. Agassiz again we were advised by him and Dr. Engelman[n] to go on with our expedition as first projected by you; and it is now our determination to start up on the company's Steam Boat on Saturday next, unless we should in the meantime [have] received instruction from you not to do so.

The company talk very fair and appear willing to give us all the necessary facilities. I have had several conversations with Mr. Culbertson who appears to be quite a gentleman. He says they will furnish us men at $25. per month. Horses at from $60. to $75. each. A suitable boat for $50. He thinks we had better take two men and two horses with carts and proceed to the locality, and after making our collection we can send one of the men back to the fort after an ox team which will be able to bring in at least two thousand pounds weight.

Evans has not got a troop of U. S. soldiers but only takes out some five or six men and will only be on the ground a few weeks, after which he starts on to Oregon and Dr. Shumard returns with their collection. Every person says the field is large enough for any quantity of collectors. Dr. Evans says he would like to see the locality well searched by as many American collectors as may wish to go there.

I must confess that the last four or five days have been passed by me under extremely unpleasant circumstances; but it has been a great relief to me to be able to avail myself of the advice of two such high-minded and distinguished gentlemen as your friends, Dr. Engelman[n] and Prof. Agassiz. Prof. A. will see you on his way to the east, and will give you a more detailed account of my course in this matter.

I delivered your message to Prof. Swallow and Dr. Litton. They have not yet received your letter and do not know your terms. Dr. Norwood is here. I have much more to say but have not time to write. I will drop you a line occasionally as we pass up the river.

In great haste I remain,

Sincerely yours, etc.

F. B. Meek[3]

The next day George Engelmann gave his version of the events in a letter to Hall:

St. Louis, May 20th, 1853.

My dear Professor

Your young men have at last left here — but you will have heard that from them several days before this reaches you, and our friend Agassiz has no doubt explained to you every thing which remained doubtful.

Your letters were all received in due time and were acted upon. Mr. Hayden as you know felt at a loss to account for your long silence. Then came Gov. Stevens & Dr. Evans and his expedition and their antagonism to your plans. Of all this you have heard already. Also that your friend (Meek) applied very earnestly for advice to Agassiz and to me. Both of us thought that the wisest plan would be to give up the expedition if you had your expenses refunded and if both young men were suitably and honorably provided for in the government expedition. After a long diplomacy (conducted by Agassiz principally) this was agreed to by Dr. Evans, but finally rejected by Gov. Stevens. We had thought that under these conditions you would release Mr. Meek of his contract, and that both of them after this season of apprenticeship could be much more useful to you. But the governor's refusal at once altered the state of things. The question was to go or not to go — and we advised them to go if they felt the courage and spirit to cope with the great advantages the others had. But whether by land or water remained doubtful. You however must have been advised of the reasons which induced them to go by water.

The last difficulty was that your or Mr. Chouteau's letter of credit did not arrive here. But Mr. Meek was so positive that you had set aside 1500 or 2000 dollars and that every thing would be arranged in a few days, he more over showed me your letters, so that I did not hesitate to give Messrs. Sarpy & Sire, the other partners of the fur company, the assurance that they would be indemnified for any advances to Messrs. Meek & Hayden up to twelve hundred dollars. With this assurance they declared themselves satisfied.

I was deeply pained to witness the spirit of rapacity, envy, and sickly emulation evinced by most of the persons interested. And of this, Mr. Hayden is not perfectly free himself. He appeared perfectly astonished when I told him that hundreds of years hence more valuable discoveries would be made in the Mauvaises Terres than now by him, Meek, Evans, Shumard and all the others together — or that the extent of the country was so vast that they might all travel and collect there without once meeting. He wanted to exclude, if possible, those German collectors of whom I had spoken, etc.

There is a work of the true spirit of Science, the pure love for Science in all things; there is on the contrary a selfishness and rapacity manifested in this, which grieved me much. And I am afraid that this is not the end of it yet.

Your Meek and Hayden however have energy and enthusiasm and will, I hope, overcome many difficulties; what the latter lacks in discretion he makes good by his candor I think.

I hope to be able to give you good news from them after the return of the boat (the Robert Campbell) which will take place in about 2 months.

Very truly, my dear sir,

Yours,

G. Engelmann

We have had a delightful time here with Agassiz, and will be glad to see him return here next spring. Is it true that you are coming shortly?

Those Germans are not going (to the Mauvaises Terres) but have gone to the interior of Missouri and from there to Arkansas & Texas principally to collect fishes for Agassiz.[4]

Engelmann also wrote to Baird about these events, in the postscript of a letter which he addressed to Baird from St. Louis on June 28, 1853:

You ask about those expeditions to the Mauvaises Terres. There are now only two: Dr. Evans of Gov. Stevens Party, with Dr. Shumard; the latter intends to return this fall, while Evans will go on. Shumard has been appointed Assistant Geologist of Missouri, by Prof. Swallow our geologist (apparently a political… but not a scientific appointment of our Governor — Swallow I mean). Shumard will reside here, he told me — he will be a valuable acquisition to us.

They go with all the means money and influence can acquire, and no doubt will do well. They went on the fur company's boat.

On the same boat went the other Party, two men sent by Prof. Hall, Meek and Hayden. Agassiz and myself tried to unite the parties — wanted Stevens to pay Hall his expenses hereto and take the two men in his employ. Evans appeared to be willing, but Stevens objected — at least to Hayden. Said to have acted in an overbearing and by no means conciliatory way; so that the two young men went out alone, without patronage and with limited means, and will, with their zeal, I hope do a great deal. Still it would have been better to have them all united. Stevens is said to have acted ungenerously — at least that is my supposition; and to have threatened that now he would throw a greater force, more men and more means in to that part of his expedition, sweep the country, etc. etc...

G. E.[5]

Thus Meek refused to be bought out by Gov. Stevens and Dr. Evans unless Hayden would be taken on board as well. Since Gov. Stevens rejected taking Hayden, Meek and Hayden kept their independence, and Dr. Engelmann paid their expenses when Hall's letter of credit failed to arrive. Meanwhile the German collectors were eased out of the picture by Agassiz, who convinced them to collect fish in Arkansas and Texas, so at last Meek and Hayden got underway to the Mauvaises Terres. The echoes of this comedy of errors reverberated for several years. It may be dismissed for the present with excerpts from three letters by James Hall, all expressive of his reaction to the outcome of his plans. Two letters were to Engelmann; the third was to Joseph Henry, Secretary of the Smithsonian Institution:

Steamboat Vienna,
Ohio River, June 15, 1853.

My Dear Dr. Engelmann,

Your letter of the 23[d] May I find in my portfolio. I had nearly finished a reply to you before leaving home, but it was left on my table. I hope Mrs. Hall will forward it to you. I did not see Agassiz on his return.

I shall attend to any drafts by the Fur Co. and feel greatly obliged for your kindness in vouching for me. I had supposed that this was all fully arranged by Mr. Cho[u]teau.

I am certainly indebted to Prof. Agassiz and yourself for your advice to Messrs. Meek and Hayden, and I have little doubt that had I been there the expedition would have been abandoned. I regret that my objects could have been misapprehended. Had I known of any intention to send an expedition then I would not have sent one. I did give Mr. Hayden some encouragement, but only in the even[t] of nothing else being done and [it was] when we all supposed that nothing would be done by the Government or any one else that the plan was attempted to be carried out. If you receive my letter you will see more fully how the matter stands. However, since I cannot recall the expedition I will hope for the best though I confess I do not expect much under the circumstances...

Prof. Swallow wrote to know if my services could be obtained in the Palaeontology of Missouri. I replied in the affirmative, but reserving the right to accept only if the conditions suited me. However, without waiting for my reply, it was given to Shumard. However it is not probable that an arrangement could have been made by which I could have taken it.

I regret that the spirit you speak of should have appeared at all and more that Mr. Hayden should have shown it. This should have been left for the parties of the other side. I hope Mr. Meek is free from such feelings, though the exhibition in others may provoke it in him. I will endeavour in future to be so cautious that I shall not crop the thach [*sic*] of others. I learn that I have a very bad name in the west as attempting to monopolise everything. I did not suppose the little I had done so much less than I ought to have done, could provoke such feelings when the field is on all sides so wide and so unoccupied.

I think hereafter I will endeavor to be content without showing my face west of the Alleghenies or south [of the] New York line!

I will send to your care a letter for Mr. Meek so soon as I have time to write him. He suggested that one could be forwarded by the Fur Co. facilities.

Under many obligations

I am very sincerely yours

Respectfully,

James Hall[6]

As regards Hall's other letter to Engelmann, the date and place of writing are not known, as only the last part of the letter is preserved. This part begins with a sentence quoted from a letter that Baird had written to Hall: "I think we have arranged matters so that the Dutchmen shall have no chances." The rest of the fragment from Hall to Engelmann reads as follows:

I received this letter on the evening of the 9th after Mr. Meek had left for St. Louis. I wrote Baird that I was much surprised and annoyed that he should have kept this information from me, that had I known it I certainly should not have sent a party into there, and that I feared difficulty. I wrote to Meek that he had my consent to abandon the expedition.

In reply to my letter Baird writes me as follows — May 12th. "My dear Professor, I see that you have clearly misunderstood the gist of my closing paragraph. By the 'Dutchmen' I did not mean your men if they be of this honored nation, but those two Germans from Europe, agents probably of Krantz,[7] who have been preparing for an incursion to the Mauvaises Terres so long. I know perfectly well that neither Dr. Evans nor Major Stevens will do anything to interfere with your party, but on the contrary, cordially assist them, and probably enable them to do more than in other ways. The field is quite wide enough for Americans, if not for foreigners. You may depend upon it, therefore, that while Major Stevens will not consider your trip an interference with his operations, his will not interfere with your expedition at all. I wrote you on Ap 2nd that Dr. Evans would probably pass through the Mauvaises Terres, on his way to Oregon. His arrangements were not fully matured until a day or two afterwards when as Geologist to Gov. Stevens expedition it was concluded to give him funds enough to make a detailed expedition."

There was not time as I supposed to write to Mr. Meek, but on receipt of his telegraphic communication of the 17th I replied that I was satisfied to abandon the expedition. This it appears he never received.

You are able to see the predicament in which I have been placed. In the first place I wished to benefit Mr. Hayden. I could expect nothing myself. Afterward I was satisfied that Hayden had not the necessary experience and then as Mr. Meek was anxious to go I sent him forward. I shall incur a large expense which I can ill afford. I have incurred the ill will and insinuations of other parties, and made myself enemies as I have well learned — and the end is not yet. I am so utterly disgusted with this whole affair that I shall make an effort to dispose of all the collection, should one be made, without even seeing it. At least this is my present feeling. I wish not to be reminded of all this trouble and annoyance to my friends as well as to myself.

Very truly yours,

James Hall

P.S. Inclosed is a letter to Mr. Meek which I would be glad to have forwarded to him the first opportunity.

Sometime about the first of May Evans telegraphed Mr. Meek asking him to go as his assistant to Oregon. Meek made no reply. I was absent from home at the time.[8]

To Joseph Henry, Hall wrote:

New York, July 23d, 1853.

Prof. Henry

Dear Sir

...In reference to the western expedition I never before heard a word in reply to my letter to you. I had written Prof. Baird twice during the winter asking whether there was any prospect of a government expedition to Nebraska. He did not reply to these inquiries, & only sometime in April did he remark incidentally that "Evans will pass through the Mauvaises Terres on his way to Oregon", and it was only in May after my expedition left and one of my party had arrived in St. Louis that I learned of the intended exploration. Had I known this I certainly should not have sent men there. I had promised Mr. Hayden during the winter conditionally, or in the event of there being no government expedition, to give him means of going there. I wrote at the same time to you on the suggestion of Dr. Leidy who had strongly sympathised with my plans, & himself fully believing that no government expedition would be sent out. Mr. Hayden had proposed remaining during the winter & was very desirous of obtaining instruments, etc.

The affair has been a very unpleasant one. I fear that the end of it will be no better than the beginning.

I am very sincerely,

Respectfully yours, etc.

James Hall[9]

The dour comments of James Hall need not sound the final note in this chapter. Rather, one may conclude it with the terse statement of John M. Clarke, who wrote in his biography of James Hall, "But the young men were off, and they harvested an experience which laid the foundation of their useful careers."[10]

6

The Trip to Fort Pierre

The steamboat *Robert Campbell* left St. Louis on May 21, 1853, loaded with cargo and carrying a quota of passengers, bound for the forts on the upper Missouri.

Although the trip up the river was, in comparison with modern travel, exceedingly difficult and slow, nevertheless it was preferable in many ways to the alternatives of going on foot, or on horseback with wagons. As a method of travel in the West, steamboats provided the ultimate in safety and comfort possible at that time.

The *Robert Campbell* could move only by day, night travel being too dangerous because of swift, treacherous currents and ever-changing channels and sandbars which could ground or wreck the boat, and also submerged snags that could rip a boat open. Frequent stops had to be made so that the deck hands could load wood onto the boat to replenish the fuel supply. Even when moving only in the daytime there were unavoidable delays, such as those due to thunder storms and to mechanical troubles on the boat. At best the *Robert Campbell* could make only a few miles an hour. The destination of Meek and Hayden, Fort Pierre, was 1,600 miles upstream from St. Louis and was not reached until June 19th, the journey to this point having consumed virtually a month.

Despite its slow pace, travel by steamboat in the 1850s was by no means wholly tedious. Those on board had ample time to get acquainted through visiting and recreation. When the boat stopped each evening, they could go ashore and gather convivially around bonfires. Often there were adventures and misadventures, spicing the trips with danger. Not all exigencies could be anticipated; some had to be dealt with at once, impromptu and decisively. Even seasoned veterans like Captain Sarpy and Alexander Culbertson, who repeated such trips year after year, could scarcely have found the trips monotonous. Newcomers like Meek and Hayden found plenty that was novel and even exhilarating as they gradually left behind the evidences of civilization and penetrated deeper into "the West." One could wish for detailed records of all the steamboat trips of that eventful period, but most were not chronicled with any fullness, and of the records that were kept, relatively few are now extant.[1] The details of the 1853 trip are historically interesting, therefore, quite apart from their bearing on the personal fortunes of F. B. Meek and F. V. Hayden.

These young men felt the usual attractions: celebrities of one kind or another on board, ranging from European nobility in search of adventure to mountain men, veterans of the fur trade who were headed for the frontier or beyond; motley, colorful groups comprising the other, less distinguished passengers, and the boat hands; raw young settlements passed in the lower valley; the chain of forts situated at strategic points along the river; interesting wildlife of many kinds; and the Indians. Meek and Hayden were alert to these novelties, and they were keenly interested also in a number of features which had less significance for the others on board. As naturalists who were intent on studying the scientific features of a little-known region, they had a lively curiosity about all aspects of the flora and fauna, the soils and meteorological conditions, the new landscapes which came into view as they progressed up the river, information gained here and there about the geological formations and the fossils which these formations contained, the traits and behavior of the Indians, and the dealings of the Indian agent with them.

With limited opportunities for close-up studies during the daytime when the boat was generally moving (or when, if it halted, it did so at points poorly suited for their investigations), Meek and Hayden tried to make the most of their evening stops. Undaunted by the difficulties presented by working at night, they showed themselves to be resourceful as well as industrious. As soon as the boat was tied up for the night, they eagerly hurried ashore, taking advantage of what daylight remained and continuing their activity after dark by the light of campfires or by using candles and lanterns. Thus they made such observations as they could, and diligently collected plants, animals, minerals, rocks, and fossils they found along the way.

Hayden's record of the 1853 journey has disappeared, presumably lost with almost all the other diaries and journals of his many years spent in the West, and no letters by him from this trip have been found. With respect to Meek's records we are more fortunate. Preserved are the diaries in which Meek made notations on the trips to and from Fort Pierre and on the excursion into the Badlands.[2] Also Meek, being now the senior partner of the expedition, wrote James Hall a number of progress reports from on board the *Robert Campbell*. The originals of these letters have disappeared also, but most if not all of them, seven letters to be exact, were published by George P. Merrill in his *The First Hundred Years of American Geology*.[3] Thus there are two accounts of the journey, both by Meek, and so far as known these are the only detailed records of the 1853 voyage. Meek's diary and his letters tell essentially the same story but they differ in many details. However, the letters provide the more readable narrative, and they are therefore reprinted here as an account of the boat trip from St. Louis to Fort Pierre in 1853:

Missouri River S. B. Robt. Campbell,
22nd May, 1853.

My dear Sir:

I intended to write you again before leaving St. Louis, but I had so many things to attend to that I had no time to do so. We left St. Louis at 12 o'clock M. yesterday and are now about sixty miles up the Missouri. Inasmuch as the company had received no letters of credit for us from the house in New York before we left we had to get Dr. Engleman[n] to go down with us and vouch for the payment by you of all our expenses at Fort Pierre as well as for our passage up and the freight on our provisions, etc. Our outfit, including our own clothing and traveling expenses, together with our boarding here, exhausted all the funds you paid us excepting about twenty-five dollars. You paid me at Albany, before I left, including the ten dollars you handed me that morning, and the ten dollars Mrs. Hall let me have a few days before, one hundred and twenty dollars; and Mr. Hayden paid me at St. Louis, of your money, one hundred and forty-five dollars and fifty cents, making in all two hundred and sixty-five dollars and fifty cents ($265.50). I keep an account of all I pay out, and Mr. Hayden has kept an account of all his own expenditures. We found upon inquiry that it would be necessary for us to have guns, and consequently purchased two at ($15.) fifteen dollars each. The gunsmith obligated himself in writing to take them back on our return at ($12.) twelve dollars each. Our expenses at the hotel were more ($18.70) (eighteen dollars and seventy cents) than we should have spent in that way. I would have taken boarding at [a] cheaper house had it not been for the fact that we had no idea of staying so long. I arrived at St. Louis on Saturday the 14th inst. and the boat was to have left on the 15th. But they postponed their departure from day to day until the 21st. We also had to purchase provision, etc., under very unfavourable circumstances, as it was only two days before the boat started that it was definitely determined that we should go. The company charge six cents per pound on all our things excepting our baggage. They charge Evans and Stevens ten cents per pound to Fort Union; and an Indian agent on board tells me he pays the company $7000 freight on the goods he now has on board for Fort Union. Our whole outfit weighs about 1900 pounds the freight upon which will amount to one hundred and fourteen dollars. I found it would be cheaper to pay even this enormous charge than to depend upon purchasing anything at the Fort, for I am informed that they sell flour there at from sixteen to eighteen cents per pound and everything else at corresponding rates.

I have had several conversations with Mr. Culbertson and others on the boat, who have been all over the Bad Lands; and they all agree that almost any quantity of bones may be collected there. They say these remains may be found over a considerable extent of country and that there is no danger of the locality being exhausted.

The Germans say that they are not going to the Bad Lands, but that they are going to Fort Union, and thence on to the Rocky Mountains and perhaps to the Pacific. One of them is the Prince of Nassau and another his traveling companion. Two others appear to be servants. The Prince is quite a youth, not more than twenty years of age; and none of them appear to have any taste for natural history.

Evans' party consists of himself, Dr. Shumard, Mr. Hendrick and an artist from St. Louis. They only expect to stay about three weeks at the Bad Lands. Shumard, Hendricks and the artist are to return, and Evans is to go on to Oregon. There are altogether some eight or ten of Stevens' party on board. Five or six of them are soldiers and the others are topographical engineers, etc. They go immediately on to Fort Union where they are to be joined by Gov. Stevens who will cross over from St. Paul with the larger part of his corps.

There is only one white lady on board. She is the wife of one of the Indian agents. The wives of several traders are on board, and all of them are squaws or half breeds. Culbertson's wife is a full blood squaw!!!

The boat has only landed three times (to take on wood) since we left St. Louis, and as there are some two hundred hired men on board they only stop a few minutes at a time, consequently Mr. Hayden and myself have had but little chance to be on shore. We made, however, good use of these few opportunities to collect botanical specimens, and have already collected some forty or fifty species. We are the only collectors on board, and the passengers turn out and help us. We have no opportunity to study the geology of the country as the boat always stops at wood yards which are generally on low alluvial bottoms some distance from the bluffs. The river sometimes runs close along under these bluffs which appear to be, and I am told are, heavy bedded Carboniferous limestone,[4] and sometimes rise to the height of 450 to 500 feet about the river. The slope or Talus is generally covered with a dense growth of oak, elm, maple, poplar, etc. whilst the rugged cliffs which tower high above are studded with stunted cedars and a few deciduous trees. The most conspicuous plants now in flower along the river banks are the anemone cylindrica, evigerou bellidifolia, senecio lobatus, etc. When we get higher up the Capt. tells us they will have to stop and cut their wood, which will give us a better chance to collect plants and fossils.

From the best estimate I can form the balance of the expense of the expedition will not cost less than a thousand dollars, and perhaps a little more. I will write perhaps once more, as we go up, and once by the boat as she comes back. After that they tell me it will be uncertain whether we will be able to communicate with our friends in the states or not. If you write direct to us at Fort Pierre, care of Pierre Cho[u]teau & Co., St. Louis.

Respectfully,

F. B. Meek

P.S. Dr. Evans wishes you to send him a copy of your chart. The enclosed ten dollars is to pay for it. Direct it as follows: Per Adams & Co.'s Express. Dr. John Evans, U. S. Geologist, care of Allan, McKinlay & Co., Oregon City, Oregon Territory.

F. B. M.

Missouri River, May 25, 1853.

My dear Sir:

I have only time to scribble you a few lines, merely to inform you that we have progressed this far on our journey, in good health and spirits. A fire, however, occurred on the boat yesterday evening which, had it not been immediately extinguished, might have blasted all our hopes of ever seeing the Mauvaises Terres. It originated amongst some barrels piled upon the bow of the boat. Some of the men had been sitting there, and had doubtless dropped fire from their tobacco pipes. Great alarm was created, because it was generally known that there were 10,000 pounds of powder in the hold of the boat immediately beneath the spot where the fire was burning. The boat was at the time under way, and nearly in the middle of the river. The men at first fled to the stern of the boat, but the coolness and courage of Capt. Sarpy soon restored order, and perhaps saved the lives of all on board.

It would be well for you to make such arrangements with the fur company as will enable us to draw enough money immediately on our arrival at St. Louis, when we come back, to pay our passage etc. from the place we take the steamboat to that city, and thence to Albany, as we will be out of funds.

Mr. Hayden would like to know if you still desire him to remain over the winter in this country.

In great haste, I remain sincerely yours,

F. B. Meek

P.S. Dr. Evans offers to lend me money enough to pay our passage down, but it may be unnecessary to borrow it if we can draw funds as soon as we arrive at St. Louis.

F. B. M.

Missouri River, May 28, 1853.

My dear Sir:

I avail myself of this opportunity to write you another hasty note. We are, as you may see, traveling at a very slow rate. The river is high and very rapid, and owing to the numerous snags and ever-shifting sandbars we have to stop during the night. Our boat even when in the best part of the river and running under the most favorable circumstances only makes about 3 ½ to 4 miles an hour.

The scenery, for some distance back, is exceedingly monotonous, as nothing is to be seen but low alluvial bottoms, and islands clothed with a dense growth of gigantic sycamores, cottonwoods, oaks, maples, etc. Once or twice yesterday evening and this morning we had a distant view of low bluffs (150 in height) which are destitute of timber excepting a few scattering clumps of scrubby oaks. This morning we passed near some low bluffs on the left, which present beautiful grassy slopes, and along the base we saw some low exposures of a buff colored rock, regularly stratified and having much the appearance of some magnesian beds we had seen in the Carboniferous limestone at several points lower down the river.

Yesterday evening we had a distant view of a Kickapoo village on a high rolling prairie; and along the river bank near the same place we saw several men and boys of this tribe nearly in a state of nudity, lazily lounging about on the green grassy sward, in the shade of some tall sycamores and spreading elms. These were the first Indians we have seen.

Last night when the boat stopped Mr. Hayden and myself lit our lantern and went ashore to collect botanical specimens, but as we had previously collected nearly every species now in flower in these low bottoms we found but one or two species new to us.

In my last I estimated the cost of the remaining portion of the expedition at about one thousand dollars. Since writing that letter I have had several conversations with Mr. Culbertson and Capt. Sarpy on the subject, and from the best estimate I can make I am inclined to think it will not be so much. They agree to furnish us six horses and three carts (able to bring away 2400 lbs.) with all necessary harness etc. for three hundred and twenty-five dollars; three men for four months at twenty-five dollars per month each, and a suitable boat for fifty dollars — so that the expenses may be approximately set down as follows:

6 horses and 3 carts, harness, etc. for three months.	$325.
3 men, four months, at $25. per month each	300.
1 boat.	50.
	$675.

This arrangement, I find, will be cheaper, and much more expeditious than to take an ox team, as I at first intended. We could probably get along while at the Bad Lands with two men, but as we would have to pay for the horses ($70.00 each) in case the Indians should steal them, I think it would be better to take a third man whose business it will be to watch them constantly. Mr. Culbertson tells us we could not come down in a boat with less than three men, one good pilot and two stout men to row. He agrees to furnish us a guide who can speak the Indian language and who is well acquainted with the country; and a pilot well acquainted with the river. We could purchase good horses at seventy dollars each, but they will not agree to take them back at the end of the season, so I find it will be cheaper to hire them.

Respectfully yours,

F. B. Meek.

Missouri River, June 2, 1853.

My dear Sir:

When I wrote you from St. Joseph I was under the impression I would not have another opportunity to address you again before the return of the boat, I have just learned, however, that we will stop at Kanesville soon, where I can mail another letter to you.

Owing to the fact that our boat generally stops where there is a wide alluvial bottom between the river and the bluffs, we have had but few opportunities to examine the geology of the country. The first drift deposits[5] of any extent noticed by us are at St. Joseph, where the hills are about 200 to 250 feet in height, and appear to be almost entirely composed of clay and sand. We have not yet seen any boulders or gravel. We stopped for the night at the base of a hill some 300 feet in height, about 15 miles from St. Joseph. On ascending this hill by candlelight we found at an elevation of some 40 feet above the river, an exposure of heavy bedded compact fragmentary limestone, eight to ten feet in thickness, in which we saw no fossils excepting a few crinoid columns. Along the steep slope, between this and the top of the hill, amongst the loose earth and broken fragments of limestone we found the following fossils, viz... [list follows]. Drs. Evans and Shumard ascended the hill at another place and found amongst the fragments of limestone on the summit, some curious... fossils...

Crinoids of Mississippian Age (Fryxell Geology Museum, Augustana College)

We have seen at several points higher up the river, low exposures along the right shores of stratified rocks, which appear to be sandstone, with beds of bluish shaly rock beneath. These rocks look like those of the coal measures,[6] but as we had no opportunity to examine them we do not know what they are.

Mr. Hayden and myself have already collected about 250 species of plants — many of them by candlelight. When the boat stops for the night the men usually make a large bright fire out in the woods, the light of which attracts thousands of insects, by which means we have collected two or three bottles full of specimens.

It is beyond the power of either pen or pencil to convey to one who has never seen similar scenery, any correct idea of the beauty of this country. The Missouri River winds about in a great valley, from 8 to 20 miles in width, which is bounded on either side by hills from 200 to 250 feet in height. The summit of these hills is the common elevation of the country after leaving the valley. The bottom land is almost perfectly level, or only very gradually sloping from the hills to the river, and is rarely cut up with sloughs. Often these bottoms and the low islands are covered with a heavy growth of trees, far over which we occasionally catch from the hurricane deck a distant view of the dim outlines of the naked bluffs. At other places these immense bottoms as well as the bluffs or hills, are destitute of trees, but thickly covered with luxuriant growth of prairie grass. But more frequently the eye wanders over a vast expanse of level prairie with here and there long, narrow belts of scattering trees and island-like groves beyond which the view is bounded by dimly seen ranges of hills, or less elevated rolling prairie. Although the prairie grass is not less than two feet high, when we look at the prairies and distant hills the whole surface appears as smooth and soft as velvet. All these objects when viewed in a proper light, present a scene which is beautiful beyond description. I am taking sketches as we go along, but these views are so vast and the objects generally so distant that the drawing has to be made on a very small scale so that it is very difficult to produce that peculiar mingling and contrast of light and shade, so pleasing to the eye, when we look upon the landscape itself.

I thought when I saw the land along the St. Peters in Minnesota I could boast of having seen the finest soil in the world. I am now convinced, however, it will not compare with that of the prairie land along the valley of the Missouri, in Nebraska and Iowa. The river banks along these bottom prairies are constantly caving in, and show a soil rich and black with organic matter as much as 6 to 8 feet below the surface. The manure the N.Y. farmer often hauls from the city to a distance of 8 or 10 miles into the country is scarcely more rich in organic matter than we here see the subsoil four or five feet below the surface. I am not speaking here of a few isolated spots, but of vast prairies, thousands of acres in extent.

We see Indians along the shores occasionally and when the boat runs close enough the passengers amuse themselves by throwing to them crackers, and pieces of bread. Though an amusing scramble takes place, I notice they always hand the bread and crackers to an old man when there is one in the party, who I suppose divides it equally amongst them. Some of these Indians are mounted on good horses and cut quite a figure as they sweep across the prairies with their long hair and trappings flying in the wind.

They say it is 1600 miles from St. Louis to Ft. Pierre so if we get safely back to Albany we will have traveled between six and seven thousand miles.

Very respectfully yours,

F. B. Meek

P. S. I have just learned that Dr. Evans is going to get off at the Council Bluffs and go up by land and leave Shumard to come on the boat. I am at a loss to know what is his object in so doing.

F. B. M.

P. S. No. 2. Dr. E. has just told me he is going to stop at the Bluffs to see about getting men and horses. He has just loaned me fifty dollars.

Missouri River, June 7, 1853.

My dear Sir:

As we are just meeting the Mackinac boats of the fur company on their way down I avail myself of the opportunity to write you once more. We are now about 30 or 40 miles above Blackbird hills, which are near a creek of the same name laid down on Owen's section of this river. On our left we have a bluff about 150 feet in height, the base of which is a soft, heavy bedded rock, having all the external appearances, as seen from the boat, of the Potsdam sandstone (No. 1 of Owen).[7] It is striped with alternate bands of yellowish brown and white, and is so friable that the bank swallows perforate it. This rock only rises 30 to 40 feet above the water and is surmounted with a heavy deposit of siliceous marl which is often seen, at various places along the river, as much as 200 to 250 feet above the water. Owen considers it the equivalent of the Loess of the Rhine.[8] This marl is often seen washed into curious ridges and conical hills which are frequently destitute of vegetation. Banvard's Panorama gives a very incorrect idea of the scenery along this part of the Missouri.[9]

On our right we have a level bottom prairie which stretches far away to the east, and is bounded by a dimly seen range of hills or bluffs of the same elevation as those on our left. The valley here, and indeed for a long distance below, is about 20 to 30 miles in width and as level as a floor. The bottoms are generally only elevated about 10 to 15 feet above the present stage of water, and are sometimes inundated. The river is constantly cutting away these bottoms and changing its course.

Dr. Evans left the boat at the Council Bluffs, and went up by land. He will meet us at Sergeants Bluff this evening, from which point he and Mr. Culbertson will go on to Fort Pierre by land with a drove of horses. They will reach that place some 4 or 5 days in advance of us. Evans says he will not leave the Fort until the boat comes, but I would not be astonished if he would push immediately on to the Bad Lands and leave Shumard to come on with his provisions. If so, he will be there a few days ahead of us. I think, however, as they only have two or three weeks to spend in that region and we can be there over three months, that we can make the most extensive collection. I did think of going up (or sending Mr. Hayden) by land from Sergeants Bluff to Fort Pierre, and thence immediately to the Mauvaises Terres. But as this would only stimulate Evans to greater exertions, we concluded we had better not do so, especially as he could employ a strong force at the Fort, if he chooses.

An interior view of Fort Pierre sketched by Alfred Sully in 1856. (Courtesy of Gilcrease Museum, Tulsa, Oklahoma)

The scenery along in this region is very fine. I am taking sketches. All on board in good health. We do not run at night, and have to stop to cut wood two or three times a day, consequently we make slow progress. The Capt. says we will reach Fort Pierre in 10 days.

Do not fail to write us to [*sic*] care of the Fur Company thru F. B. Meek or F. V. Hayden, Fort Pierre, Nebraska, care of P. Cho[u]teau & Co., St. Louis, Missouri.

Respectfully,

F. B. Meek.

June 11th, 1853.
30 miles above mouth Vermilion River.

When I commenced this letter I was under the impression that some piles of flood wood seen at a long distance ahead of us were the Mackinac boats on their way down. We have not yet met them, but are expecting to do so very soon. Dr. Evans came on board again at Sergeants Bluff, but will start from some point (I forget the name) which we will probably reach some time today. The small streams are all very much swollen so that they will not be able to travel fast enough to beat the boat up [by] more than three days. Mr. Culbertson will also go up by land, and has agreed to have our horses, carts, men and all in readiness by the time we reach F. P.

Maj. Vaughan, an Indian agent who has on board eight thousand dollars worth of presents for the Sioux Indians, near Fort Pierre, agrees to warn them not to molest us at the Bad Lands. He says he will tell them the Great Father has given us leave to travel in their country, and if they steal our horses or give us any trouble he will send them no more presents.

We stopped yesterday at the mouth of Vermilion River for Maj. Vaughan to distribute presents to a band of Yankton Sioux. As the boat approached they all assembled on the bank, men, women and children, together with a troop of gaunt, wolfish looking dogs. The men were painted and decked in their best, and were really fine looking fellows. As soon as the boat came near the shore they ranged themselves along the bank and fired a salute of several rounds. Maj. V. went ashore and was introduced to them by the interpreter and at the same time presented them, in the name of the Great Father, with some sacks of beans, flour etc., together with a keg of powder and a bag of ½ oz. balls. There was no chief present, but an old fellow came forward and delivered a short harangue, the substance of which was that they were glad to see their Father (the Agt.). They had always been poor, and needy, and were never more so than at present. They wished their father to tell the Great Father they were thankful for the present he had sent them. They were glad to see he had not forgotten them, etc. At the end of this harangue the Braves all responded How! How! Maj. V. told them the Great Father was their friend and would never forget them — that he was anxious to keep liquor dealers and bad men out of their country, and that if they would drive all such bad white men away the Great Father would be glad, etc. To this they responded How! How! This ceremony occupied but a few minutes and we soon started, leaving our red friends preparing for a great feast.

Fort Pierre, looking south sketched by Alfred Sully in 1856. (Courtesy of Gilcrease Museum, Tulsa, Oklahoma)

This morning we killed a fine elk (Cervus canadensis, Say) in the river. It was not far from shore, and could have easily made its escape, but the noble animal lingered for its young fawn which was so frightened that it could scarcely swim and as the boat approached a dozen or more rifles were leveled upon her. All fired about the same time, so that no one could tell who was the lucky marksman. The poor animal struggled for a moment and floated off.

The yawl was lowered and she was soon taken up. The young one could have escaped but it continued to swim around the dead body of its Dame until it was caught.

We are all in good health. We have collected about 350 species of plants, some of which are new to us.

Yours respectfully,

F. B. Meek

The Prince of Nassau started by land from Dorrian's Hills, leaving his men aboard.

June 12. Dr. Evans did not leave at Dorrian's Hills, and is going up all the way by boat.

Fort Pierre, June 19, 1853.

My dear Sir:

The Makinac boats have not yet started, but will do so in a few days. We arrived here this morning at 7 oclock and our things are now on shore, but owing to the fact that all the teams and men are engaged in hauling up the goods of the Company, we will probably not be able to open our boxes until tomorrow. We were within two miles of this place last night before dark, but were compelled to tie up by a tornado which came very near sinking the boat. The Indians, or at least some bands of them, are not very well disposed towards the whites at this time. Some of them do not come in to meet the Agent, and refuse to accept their annuities. They have sent word to the Company that they will not allow the boat to go higher up than Ft. Clark, and that they will not allow Gov. Stevens' party to pass through their country. Our men, horses, carts, etc. will all be ready by tomorrow or the day after. Our things will be taken to the fort this evening when we will go immediately to work to separating what we are going to take to the Bad Lands from what we expect to use on our way down. Two of our men are good guides and interpreters. One of them is a half-breed who was raised amongst the Indians and is said to be better acquainted with their habits and customs than any other person in the country. He has hunted all over the Bad Lands. He will take two of his own horses and his squaw along, and Culbertson says if we have any fighting to do he will be the last man to leave us.

Drs. Evans and Shumard will start about the same time we do and have expressed a desire to have us camp near them during their stay at the Bad Lands. They say deer, buffalo, antelope, elk and mountain sheep are very abundant there and that our guide can kill more meat than our party can use, though I do not think it prudent to rely upon this means of supply.

I hope you will excuse this incoherent letter for I am in a great hurry and have to write on a table in the cabin with about two hundred Indian chiefs and braves seated in rows on each side of me. They have come on board by invitation from the captain to a feast. They are elegantly dressed and their bearing is noble and dignified. One old fellow has just presented a fine buffalo robe to Capt. Sarpy. He first spread it down on the floor and made the Capt. sit down upon it, when he commenced a long speech which he wound up by presenting the robe, and telling the Capt. that he looked upon him as only a little inferior to the Great Father. They have given us all an invitation to a dog feast to-night. I would like to go but will not have time. The Capt., Dr. Evans & Maj. Vaughan, the Agt. will go, and they say they are going to eat some of the dog. I do not envy them their supper.

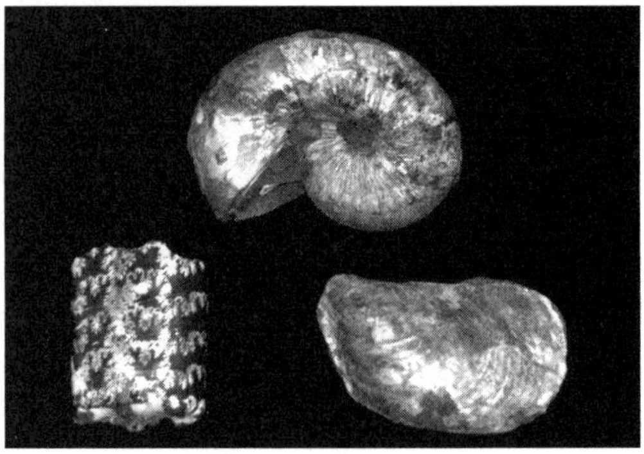

Cretaceous invertebrate fossils; clock-wise from lower left: *Baculites, Scaphites, Inoceramus.* The *Baculites* is 10 cm (4 inches) in length (Fryxell Geology Museum, Augustana College; photograph by Erwin Weber)

I have given the Company a draft on you for the amt. of the freight on our goods and for our passage, as follows:

Freight on 1871 lb. at 6¢ per lb.	$112.26
Passage for Mr. Hayden & myself ($50. each)	100.00
	$212.26

In great haste, yours,

F. B. Meek

P. S. We walked across the great bend and collected some fine cretaceous fossils — Baculites, Inoceramus, Circullea, Ammonites, Fusus, etc.

June 20th. We have all our things ready to start. Dr. Evans is not quite ready but will be by tomorrow. We have agreed to wait for him. I am much pleased with our men.

7

Fossil Hunting in the Badlands

Having unloaded freight and a number of passengers, including Meek, Hayden, and the members of Evans' party, the *Robert Campbell* with lightened load resumed its journey upriver toward Fort Union.

As mentioned in Meek's letter of June 19th, the scientists lost no time in making preparations for their excursion into the Badlands, which lay in the country to the west of the fort. Such rivalry as had hitherto existed between the two parties now gave way to friendly forbearance and mutual assistance. Both parties were small and they faced problems better solved through cooperation than by wholly independent action; furthermore their members were, after all, basically generous and kindly men. Of immediate concern was the hiring of satisfactory guides and helpers, and retaining their services; arranging for overland transportation into the Badlands and back again; and, particularly, coping with the difficulties and risks they could expect to encounter once they left behind the haven of Fort Pierre. For they would go into country where the supplies of food and potable water were not always dependable, and where they must manage without benefit of protection other than that which their own limited numbers provided.

Meek's diary consists mostly of scientific notes but interspersed with these are observations concerning the country traversed, accounts of their experiences, and comments reflecting his personal impressions. Meek's timidity, coupled with his physical limitations, may have led him to magnify the dangers to which they were exposed, but undoubtedly the hostility of the Indians was a real threat to their safety, one that made the accomplishment of their mission more difficult than anticipated. Consequently they were away from Fort Pierre less than a month, a period much shorter than that contemplated at the outset. Nevertheless, as shown by the following excerpts from Meek's diary, they succeeded in making an impressive collection, and they managed to get the better part of it back to Fort Pierre for shipment downriver to St. Louis.

Monday, June 20th

Spent the day in getting our things ready to start out. We are informed that the Indians are growing very jealous of the whites, and do not wish them to come into their country. Many of them refuse to come in and receive their annuities. The half breeds have made them believe the president wishes to get them to accept the goods and will then claim their country. The Agt. had a long conference with some of the more friendly of them at the Fort today. They all remonstrated against having the whites come into their country. They are all jealous of the Fur Co. and think the Agt. gives them [the Co.] the goods (or a large portion of them) sent by the government and that they make them [the Indians] pay for what legitimately belongs to them.

View south from Saddle Pass Trail, Badlands National Park. The Pine Ridge forms the distant horizon. (Photograph by Richard C. Anderson)

A great ball at the Fort tonight. Ladies all squaws or half breeds. Ended in a general row at about 3.oclk A. M.

Tuesday, June 21st

MAUVAISES TERRES.

All things being ready at 3.oclk P. M. we start with the intention of camping at Willow Creek, 8. miles from the Fort. Indians told our men we will get into trouble — are very much opposed to our going out — say they will kill our horses and rob or perhaps kill us.

Dr. Evans's party or at least his men with three carts (2. horses each) and a four horse wagon with two men on horseback started at same time.

The Dr. and Dr. Shumard together with Mr. Hendrix will come on and join us in the morning. My party consists of myself, Mr. Hayden and three men. Their names are Henry Clermont, Francis Rondell and Isadore Daurion a quadroon who is as dark coloured as an Indian. Has always lived with them. He takes his own wigwam and his squaw with him, also his own horse. Is said to be a fine guide and hunter and is well acquainted with the Indians and their habits. Both the other men are good interpreters, hunters, and guides. Daurion speaks French and Sioux very well but not a word of English.

After leaving the fort we traveled over a nearly level bottom prairie, due west for about a mile, when we came to more elevated prairie country. Ascent very gradual until we reached the summit elevation of the country which is about 100 to 150 feet above the river. Our road continued nearly due west over a rather hilly prairie country where there is not a solitary tree or shrub to be seen as far as the eye can reach in every direction. When we looked at the country at some distance off there seemed to be plenty of grass everywhere, but we found a great number of spots both on the hills and hill sides as well as in the little valleys where there is not a single blade of grass nor any other kind of vegetation. The soil or earth in these places and all along the road is dark coloured and one would suppose by merely looking at it that it would be productive. Along some of the slopes and ravines a kind of dark gray clayey marl[1] with seams of argillaceous carbonate of Iron was seen. As this marl disintegrates these seams are left on the surface and break into small angular pieces which are often seen so thickly over the surface as to almost hide the soil. They become very hard and turn quite black by exposure (perhaps from presence of Manganese).[2]

At a distance of about 8. miles nearly W. of the Fort we descended by winding amongst the little ravines and gentle slopes into the valley of Willow Creek which is only about 60 or 70 feet below the common elevation of the country. It is but a small stream, about 10. feet wide and 2. feet deep & runs to the S.E.

In crossing this creek two of our carts stuck fast in the mud and in attempting to pull out broke one of the shafts. Had to unload and to make matters worse there came up a hard rain. About the time we got out Dr. Evans's 4. horse team attempted to cross and got mired. The horses [mules] plunged about and fell down in the mud & water and it was at first feared they would drown, but the men soon ungeared them and got them out, soon after which they went back and camped on a little elevation. As soon as we were fairly out of the mud and loaded up we drove on a short distance and pitched our tent in a hard rain. Men sullen and in a bad humor. Started Daurion & Mr. Hayden immediately back to the fort with the broken cart with instructions to get another. Daurion's squaw hunted up a few dry branches and built a fire and soon had us a warm cup of tea, some jerked buffalo meat, pork, etc. At about 11.oclk there sprang up a very hard wind from N.W. At about 12 oclk Mr. Hayden & Daurion returned with the cart from the Fort.

While men were getting breakfast we went down to the creek to examine an exposure of the strata that underlie the country...

[Geological and botanical notes.]

Wed., June 22[d]

As we wished to wait for Dr. Evans and Shumard, Mr. Hendricks & Dr. E's artist who are to come out on horseback we concluded to wait a little for the wind to dry up the mud on the hills, where it is always very slippery after rain.

At about 11. oclk A.M. Dr. E and his comrades arrived and we started. Dr. Evans had the kindness to lead out for me a fine riding pony sent to me by Mr. Galpin as a matter of courtesy.

The country passed over today is entirely destitute of trees or shrubs, but supports a tolerable growth of grass with quite a number of plants bearing handsome flowers. The whole country as far as the eye can reach is rather more hilly than what is usually termed a rolling prairie, though the slopes are very gentle. Saw a few boulders (only 3.) not over 2 feet in diameter of redish gneiss.[3] Had several showers during the day. Our course was a little S. of W...

At about 4.oclk, came to a small stream similar to that at which we camped last night. Both tributaries of Titon or the Little Missouri.[4] Water clear and good. Found an old half breed by the name of Pino and an Indian together with their squaws encamped here. They were just returning from a beaver hunt on L'eau qui Court.[5] They had caught 56 beaver. We purchased from them some fine venison. They tell us there are about 300 lodges of Sioux encamped near the Bad Lands who say they will rob and drive away any white men who may come into their country.

At about sun set the Indian sent us an invitation to a feast. On arriving at his lodge we found a tin pan full of jerked buffalo meat & pome de prairie boiled together. The dish did not look very clean nor was it very savoury, but as it would have been a great breach of etiquet [*sic*] to have appeared displeased with it, we eat of it and pronounced it very fine. The lodge looked comparatively neat & comfortable. The Indian and his squaw did not [look neat and comfortable] but the former conversed freely and appeared to be pleased to see us eat.

Thursday, 23d June.

Started by sun rise and came on to Lance Creek 12. m. to breakfast. Country similar to that seen yesterday. Arrived at Lance Creek at 9 ½. oclk. Nearly cool enough for frost last night. Horses stalled again & had a good deal of difficulty in getting over the mud in creek. Came on after breakfast to a small stream called Sipi Creek (15 m.) where we arrived at 4 oclk to camp for the night. These early starts & stops give the animals time to feed. Country traveled over today not quite so hilly but still rolling and undulating, no trees or shrubs of any size to be seen excepting a few small cotton woods and willows along small streams or rather shallow valleys with very gentle slopes, in which there are streams of water after rains and at this time, excepting in one or two, only pools of water. It is clear and cool but does not taste very well. Geological characters of country similar to that already seen between here & Ft. Pierre... Had one or two showers of rain during the day, and quite a shower soon after we camped — followed by sunshine. Saw an antelope on the prairie today but it was very wild and ran away before the men could get within gun shot of it. It was first seen and pointed out to the hunters by Dr. Evans's scout, an old Indian Medicine Man who rides ahead in order to apprize us of the presence of game or Indians. He is quite a dignified old fellow. His name is, in Sioux, Esonni Monza (Iron Horn).

Saw quite a number of plants in flower along the road today... Not near so much wind today. Passed by several prairie dog villages.

Friday, June 24th.

Started before sun rise and came on to Reeare Creek 9.m to breakfast. Morning cool but clear. Temp. 42°-. Arrived at breakfasting place at 8 ½ oclk A. M. This is largest stream we have seen. No running water but holes 10 feet deep. Water clear and cool. Only a very few small willows scattered along the margin. Isadore took my riding pony and rode out to East of us to see if he could not kill some game. We saw an antelope as we came along. Same plants seen yesterday continue today... Geological & physical features similar to that passed over yesterday. I. E. rolling prairie, with a difference of about 90 to 100 feet between lowest depressions and highest elevations. Slopes very gentle... One of men killed a prairie dog. Passed several of their villages. They are very shy; always run to their burrows which enter the ground nearly vertical in the midst of a small hillock of earth. When wounded they nearly always run into their burrows. Men cooked the prairie dog for breakfast — it was very fat and tasted only tolerably good, though the men and Indians think it very fine.

Country becomes gradually more elevated and less hilly. Horses did very well today.

At 3.oclk this afternoon Isadore joined us with two fine antelopes he had killed since morning. At 5.oclk arrived at Big Cottonwood Creek where we camped for the night. Had quite a feast of antelope meat. I am very fond of it, think it more savory and delicate food than deer meat. The men say they cannot undergo so much fatigue when confined to it as when they have buffalo meat. Rondell took the tripe as soon as Isadore deposited his load on the wagon, and ate it raw without washing. Appeared to think it a delicate morsel. About the time we were fixing to camp, Old Iron Horn, Dr. Evans's guide & hunter, came in with two more antelopes.

After supper we walked a short distance up the creek where it has cut in the side of a hill exposing the following section:

[Geological notes.]

Sat., June 25th.

Started at sun rise. Sun rose clear but soon clouded up. East or S.E. wind cool & unpleasant. Soon after leaving the shallow depression where we camped for the night, came in sight of Grindstone Buttes. Country & vegetation exactly similar to that passed over yesterday. Breakfasted at a small creek within about 2 miles of Grindstone Buttes, Grindstone Creek, & 12. miles from our camping place last night. Only a few small willow & cottonwood trees along stream. Saw antelope on prairie. Several trees six to 8 in. in diameter have been cut down here by beaver recently.

After breakfast rode on to Grindstone Buttes which was nearly west of us. After riding over an undulating prairie about two miles came to the base of the range of hills, which rise not less than two hundred feet above the common elevation of the country over which we have been traveling from breakfasting place to the Butte. There is also an exceedingly gentle rise in the surface from the small stream where we dined to the base of these hills (about 100 feet in the dist. of 2. miles)... Buttes appeared to range nearly N.W. & S.E. Immediately to W. or S.W. lower country, but at a dist. of 1½ or 2 miles S.W. a parallel range of hills of less elevation, far beyond which there is another parallel range which appear as seen in distance to be rocky — peaks of Bad Lands. All round to S.E. — E. & N.E. country lower and approximately similar to that over which we have been traveling. View vast and beautiful but lonely. Not a tree or shrub in sight excepting a few scattering willows & poplars of diminutive size, along the little stream where we breakfasted. Off to N. a few isolated round hills which rise to near the elevation of the other Buttes.

These Buttes are entirely isolated from any other high country. Some pebbles of dark flint or horn stone embedded in sandstone. Several piles of loose stones on top of Buttes evidently placed there by hands of men. Probably Indian graves, or signs of some sort…

Crossed a damp place or rather a pond full of grass… in which we saw great numbers of Garter Snakes, some of which we caught for preservation.

Camped for the night on a small creek which is one of the head branches of Little Missouri or Titon River. Whole dist. today about 30 miles. Shower of rain with thunder & lightning. Great numbers Astragalis & other leguminous plants today.

Sun., June 26.

After leaving camping place this morning ascended a very gradual [slope] until we reached a plateau probably 80 feet above creek above where we camped…

After this hasty examination mounted my horse and rode down the hill to the trail and followed on after the carts, which by this time were some dist. ahead. The trail continued on up the valley of a small stream for some distance with a very gentle ascent and finally turned a little to the right and ascended a tolerable steep ridge. When I was fairly on the summit I found myself on a nearly level plateau. The scene that now presented itself was grand and beautiful. Directly in front and in a S.W. direction were to be seen looming up above the horizon & as it were from the level prairie the dark blue peaks of the Black Hills, with white places which I afterwards learned were snow. They were distant about 75. miles and looked gloomy & grand. Looking off to the S. at a distance of 10 miles I caught the first view of the long looked for and far famed Mauvaises Terres. The appearance was very much like that of the ruins of Ancient Castles and fortifications of gigantic dimentions. Although there were several valleys between me and these peaks, when seen from my position they appeared to rise from the level prairie. A little west of those first seen is a long ridge of the same elevation & doubtless of same formation, the eastern part of which is cut and worn by the rains into singular fantastic forms of towers, etc. These hills & peaks appear to be entirely destitute of vegetation, and are light grayish yellow, almost white in col.

Away off to the west I saw a line of dark spots which I knew was our horses & carts. After riding about 4 or 5 miles in a west course, all of which dist. [was] down an exceedingly gentle slope, so much so as to be almost imperceptible, I ascended a hill which was about 60 feet in height, the summit of which I suppose to be nearly upon a level with highest part of the prairie after leaving the valley of the head stream of the Little Missouri. Along the brow and on the level summit of this hill saw many loose flat pieces of white hard limestone and thicker pieces flint or chert. After riding a short dist. I caught up with carts & we descended from this hill to a spring called Pino's Spring where we halted to breakfast, having traveled 18 miles this morning…

Isadore started out hunting this morning soon after day light and joined us at our breakfast halt with a fine antelope. Saw another feeding on hill near us while we were eating breakfast. Old Iron Horn killed one this morning & another of Dr. Evans's men killed another this evening in the road only a short distance ahead of us.

After leaving Pino's Springs we went up a very gradual ascent upon a nearly level prairie where the ground is very dry and hard. Soil light col. Only scant growth of grass. Ground cracked and dry. Many cactus... Continued our journey over this prairie about 10 miles to Bull Creek, a tributary of the Chayenne — the first stream we have seen on the route which runs in that direction. Our course all the way since leaving Pino's Spring and for some dist. on other side has been on dividing country between Little Missouri & Chayenne. All the way from Pino's Springs to our camping place had peaks of Bad Lands on our left and gentle swell on our right which frequently intervened between us and Black Hills. When within about 3. miles of Bull Creek our trail run close to some of the breaks of Bad Lands. I rode over to examine them. Found on approaching nearest peaks a shallow valley or ravine between me and them. Descending this saw near the brow and scattered down the slope, thin flat pieces of white hard limestone in which I found some few specimens of a small univalve. Slopes of hills washed into thousands of little ravines. Hills as far as examined appeared to be wholly composed of white & light redish clay or marl with thin seams of silicious limestone? As it was late and there was a thunderstorm coming up I was obliged to hurry on without making further examinations.

Bull Creek has cut a bed in the strata at the place where we camped, which shows the following section (This place is about 3 or 4 m. west of the continuation of the Bad Lands seen on our left all day):

[Geological notes.]

Isadore came in towards night with several fragments of bones which says he found on the top of one of the peaks in the Bad Lands.

Mon., June 27.

Started at sun rise and came on to Sage Creek 12 m. to breakfast, country all along more broken and hilly. As [we] approach Sage Creek becomes much more so. Descending into Sage Creek Valley, found the creek at this time having but very little running water but some deep pools. Caught a few small fish. Water clear but unpleasant to drink, tastes like epsom salts. Creek very crooked in its valley, and has cut a bed 10 to 15 feet deep in a kind of drift or alluvial deposit of sand and gravel.

Veins of chalcedony (finely crystalline quartz) formed in cracks in the Badlands sediments. Being harder than the enclosing sediments, they weather into relief, forming wall-like structures such as those casting prominent shadows in the lower center of this photo. (Photograph by Richard C. Anderson)

Rapidly eroding badlands ravine; Shalimar Ranch, Harrison, Nebraska (Photograph by Richard C. Anderson)

Camped near the forks on West Fork. Below camp N. 20°W. country very hilly. S.E. up creek much more so, and high naked peaks of Bad Lands in the dist. S.E. bound the view. Scenery remarkable and peculiar. Took sketch. About ½ a mile N. 20°W. of us the creek has cut close in under the hills on the left hand side where the following section is exposed:

[Geological notes.]

Along in the bed of creek we found by breaking open nodules fine specimens of Ammonite, Baculite, Scaphite, Inoceramus, Nautilus, Dentatum etc. together with some small univalves.[6] All these fossils are found in the lower clay beds but they are generally crushed, and fall to fragments when the crumbling clay is touched. Some of the Ammonites are very large, in nodules, not less than 3. ft in diam.

As we think it unnecessary to haul our specimens to the end of our journey and back we have made a cache of them here, and intend to get them on our return. This is certainly a very fine locality. Two or three men could almost load a cart with fossils in two days.

Old Iron Horn, Dr. Evans's guide & hunter, advises us to go back and appears unwilling to go any further with us. Says we will get into trouble with the Indians.

Some dissatisfaction amongst our men almost amounting to mutiny. Have succeeded in getting them to go on.

Water of creek acts on my bowels. Much white shrubby plant with strong camphor odour (Artimesia). Called wild sage, hence name of creek. Abundant.

Tuesday, June 28th.

Being all very much fatigued by yesterday's labour and having our fossils to wrap up & hide as well as our notes to write up, did not start until nine oclk A. M. After which we ascended from the valley of Sage Creek upon an elevated plateau (200 to 250 feet) where we had fine view of Bad Lands on our left and of Black Hills on our right, or rather nearly in front of us. We are now only about 30 miles from Blk. Hills the nearest point on our route. Caught a rattle snake this morning. At about 12. oclk M. arrived at Bear Creek where we stopped for dinner. Same geological formations here as at Sage Creek, but not so good a locality for Cretaceous fossils.

From our camp which is in the valley of the creek S.E. up the creek [we] can see the peaks and buttes of the Mauvaises Terres which are only a short distance, and even the hills on each side of us and for some distance to W. & N.W. appear to be capped with the Tertiary clays. After dinner went up to the Palaeotherium beds.[7] Collected some fragments of bones; find them mostly fractured and so firmly embedded in the indurated clay that it is difficult if not impossible to extract an entire jaw or any of the long bones. Generally even the teeth are fractured and come out in fragments.

I have heard a great variety of descriptions given of the Mauvaises Terres, none of which had conveyed to my mind a very correct idea of them. At this particular locality the following is as nearly a correct description as I can give.

Our camp is as I before said in the valley of Bear Creek, a small tributary of the Chayenne. This little stream, which runs about N.30°W., has at this particular spot cut its way down through the Tertiary & Cretaceous rocks, forming a valley about 150 feet deep, the bottom of which is only about two hundred yards in width. Hills or bluffs [have] gentle slopes. In this alluvial bottom, after leaving the Bad Lands, the creek has cut a crooked course about 50 to 60 feet wide and 10 to 15 feet deep with nearly perpendicular banks of hard white clay. At this time there is but very little running water in the creek, which is strongly impregnated with something that tastes like sulphate of magnesium and acts upon the bowels as a purgative.

At a distance of half a mile (S.E.) up the creek there is more elevated bottom, 70 or 90 feet above the other, which widens out into a kind of amphitheater about 8 to 10 m. in diameter. In this great basin the little creek on which we are camped, & its branches & another smaller stream, which is also a tributary of the Chayenne, have cut an endless labyrinth of ravines and flat depressions, nearly all of which are now dry. Only have pools of brackish water. Between these ravines there are frequently isolated patches of ground nearly all of the same level and overgrown with a scant growth of grass. Also a good deal of cactus… wild sage, Artemesia, etc. Often the spaces between the patches of elevated grassy ground are as wide, or wider than the patches themselves, and present a nearly level or only gently undulating surface (with small ravines), as white almost as snow and destitute of vegetation. The surface of these lower barren spaces are nearly as hard as rock & often covered with fragments of thin translucent quartz or calcedony which rattle like pottery under the horses' feet. I was at a loss at first to know the origin of these fragments but soon discovered they occur in veins which traverse the Tertiary clays in almost every direction. I think it was mentioned in Culbertson's journal that there were veins of selenite.[8] In looking across this great amphitheater, especially when the observer is at a certain elevation, its bottom looks like an almost unbroken grassy plain, but in attempting to cross it the traveler finds himself compelled to cross an almost endless number of ravines, in which he goes out of sight, and flat depressions where there is not a spear of grass. The isolated patches of grass are generally elevated about 2. to 6. feet above the depressions, and have generally very steep banks. Often, these more elevated patches have been worn away by the rains and waters until there is only a small island like mass left in the midst of a wide depression. Sometimes these masses are reduced to a mere white column with a little crown of grass or cactus…

In various parts of the aforementioned great amphitheatre, there are isolated conical & other formed hills 20 to 100 feet high entirely destitute of grass or any kind of vegetation. Their sides and slopes are cut and worn by the rains into thousands of little ravines & trenches so as to present a most extraordinary appearance. These hills are of various elevations from only a few feet above that of the aforementioned grassy patches to that of the highest parts of the country bounding the amphitheatre. Sometimes they are very large, level on top, and covered with a growth of same grass seen on the surrounding elevated prairie country bounding the Bad Lands. The Bad Lands proper, or at any rate those at this locality, are bounded on every side by bluffs about 250 feet in height. The descent is generally very steep and the slopes are cut into all manner of fantastic shaped ridges & conical hills, which are nearly white and often present as seen at a distance the appearance of the ruins of ancient castles, churches etc. One is more impressed with the reality of these castelated forms when the Bad Lands are seen at a distance from the high surrounding prairie country, where the peaks and long wall-like ridges appear to rise as it were from the surface of the grassy prairie.

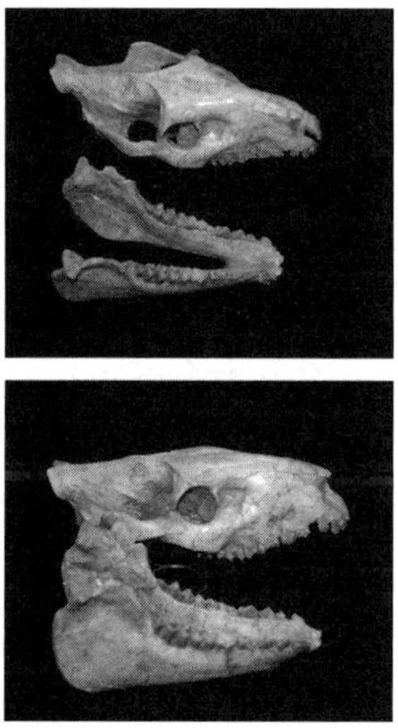

Skull and lower jaw of an Oreodont, the most common fossil mammal in the Badlands. Length of this specimen is 20 cm (8 inches). (Fryxell Geology Museum, Augustana College; photograph by Erwin Weber)

I have heard it said that these peaks & isolated hills in the Bad Lands look as though the surrounding depressions had been caused by the sinking away of the surface, and I have even heard it hinted that volcanic agencies or at least the burning out of lignite beds may have had something to do with the production of these remarkable appearances. This, however, I think is not the case. There is nothing to warrant such a conclusion for we find the rocks with few local exceptions laying [*sic*] in an undisturbed nearly horizontal (general dip 3 or 4° S. E.) position. The same beds are seen not only in the same positions in opposite sides of the great amphitheatre but in the high isolated hills between. No lava or other volcanic products are any where to be seen.

These depressions with their isolated peaks & ridges as well as numerous valleys & ravines are clearly attributable to the wearing away of the Tertiary clays by atmospheric agencies. There is always one or more small streams which head in them and as the Tertiary clays disintegrate more rapidly than the Cretaceous, the valley always becomes comparatively narrow where the stream passes from the former to the latter. These Bad Lands here on the head of Bear Creek are situated at the dividing elevation between White and the Chayenne. There is at least one (if not more) small stream which heads in them that flows into White River.

I have been too much occupied in collecting specimens to take any sections yet.

Wed. 29, Thu. 30, Fri. 1st July, Sat. 2d, Sun. 3d.

All spent in collecting bones. We have now about 700 lb. of Palaeotherium bones, mostly pieces of long bones with their articulations and pieces of jaws with teeth, etc. We have also some 20. or 25 good heads of smaller animals together with quite a number of pieces of heads, jaws, teeth, long bones, etc. of same. Weather excessively warm. Men & horses exhausted.

Have not only the direct rays of the sun but reflected light and heat from the white indurated clay. Horses' feet very sore, some of them lame. Concluded to give the men holiday tomorrow. Rest horses, etc. Saw Dr. Evans's party in field today but not in speaking distance. Found some very large turtles and fine heads today.

July 4th, Monday.

Was roused this morning by the firing of guns by the men in honour of Independence Day. Flies and musquitoes very troublesome; fear they will kill our horses. Isadore went over on the Chayenne yesterday hunting. Killed a deer but could not find it. Saw trail where large body of Indians had passed.

Our camping place and the locality where we are at work is between two large bands of unfriendly Indians, one on the Chayenne, the other on White River. We are constantly in dread of them. If they find us they will probably not kill us but will rob us of everything.

Men discontented and want to go back to the fort. We see many wolves every day. They prowl around within a 100 yds of our camp soon after sun set, and at night come up and help themselves to pieces of fresh meat left hanging on the carts within two or three steps of the tents, see tracks of grizzly bears in Bad Lands. Isadore has killed several antelopes and a blk. tailed deer since we came here, but no buffalo. A few big horn (Ovis Montana) have been seen but none killed. One of the men killed a badger (Meles Labradoria) and we occasionally see in the Bad Lands a common rabbit, (Lepus Sylvaticus?).

Tuesday, July 5.

Day excessively warm. Was unwell but spent the day in boiling specimens in glue water, packing them, etc. Mr. Hayden & Rondell went out and collected more good specimens. Dr. Evans and his party returned from their camping place over on a small creek in the Bad Lands which runs into White River, this evening.

Wed., July 6th.

Had very windy night. Morning very warm. Men discontented. One of them (Isadore) wants to return to the Fort. Says the Indians will kill him if they find him here. Have succeeded in prevailing upon him to remain ten days longer with us. Dr. Evans and his party started back to the Fort this morning.

Mr. Hayden, Rondell & Clermond[t] started out this morning to collect more specimens, and I went up the creek to take a section of the beds exposed, which I found as follows in the ascending order:

[Geological notes.]

This evening one of the horses fell over a precipitous bank of the creek about 30. ft. and soon after died. Heard a gun fire over towards White River. Men fear it is Indians coming this way, and load up their guns.

Thu., July 7.

Thunder & lightning last [night?] with wind, but only a few large drops of rain. Mr. Hayden, Rondell and Clermont started out after breakfast hunting specimens, and I rode across the Bad Lands about 7. mi. to continue my section.

[Geological notes.]

It is very difficult if not impossible to represent in a single section all the beds composing the Mauvaises Terres, as many of them are only local, whilst others that are at some places only represented by mere seams that would scarcely attract attention swell out 3 to 10 feet and sometimes more in a short distance.

As I returned to camp I observed along the bank of the creek about a mile or 2 up in Mauvaises T. that the beds below the Palaeotherium bed appear to be broken and more or less disturbed and present a brick red appearance as though they had been subjected to heat.[9] These appearances seem to favour the opinion which Dr. Evans & others expressed that the great depression in the surface of the country throughout the Mauvaises Terres has been caused by the burning out of lignite beds. Although depressions in the surface of the country may have been made in this way, and this cause may have contributed to some little extent in forming the remarkable appearances at this place I cannot for a moment believe otherwise than that it is almost wholly the work of denudation…

Friday, July 8th.

Spent the day in gluing, wrapping, and packing specimens. During afternoon two buffalo came on the brow of the hill within 150 yards of the camp and looked down at us a while, and then walked out of sight behind a hill. Clermont & Isadore our two best hunters were out hunting in the Bad Lands, but Rondell snatched up his gun and started in pursuit of them. Soon after he was out of sight I heard his gun fire and hoped he had killed one, but was soon after disappointed by seeing him return without any meat. He had wounded one but could not catch it. Our young Indian took his bow and arrows, mounted a horse, and started in pursuit of them, but they run up amongst the ravines in the Bad Lands, and he could not overtake them. Towards evening Clermon[t] and Isadore came back with a fine antelope. They had also wounded a big horn but lost it. They also killed a hare and saw many small gray ground squirrels in the Bad Lands. I have often seen these little squirrels. They are about the size of the little striped squirrels of the middle states, but are light gray and I thought I could see faint darker stripes down their back. I have never seen them excepting amongst the dry hills and ravines of the Bad Lands, & have not been able to kill one. I also saw in several parts of the Bad Lands a large hare, probably Lepus artemisia. It appears to be very wild.

Sat., July 9th.

Rode about 9. miles up in the Bad Lands today to get a sketch of some characteristic scenery. Mr. Hayden & Rondell went along and collected three good heads, some turtles and small bones while I was sketching. At this place there are several isolated hills (Buttes) which partially separate the great basin in which Bear Creek & another tributary of the Chayenne from some small tributaries of White River. The bottom of the basin which [is] not more than 1 ½ or 2. m. in width here, is the turtle and bone bed, and as there are often spaces of many acres perfectly denuded of vegetation and soil, hundreds of broken up and decomposing turtles are to be seen strewn over the surface. There were also some good ones, and good bones of other animals. As this bed extends and is well exposed across the valley and can be seen to pass under the base of the main bluffs on each side, the idea that this great valley or depression was formed by a sinking away of the surface is unquestionably erroneous as far as this part of it is concerned, and it can be demonstrated that this part of it is due to denudation. I can see no necessity for calling in the agency [of] burning lignite bed. I saw at several places where the surface of this and other beds were laid bare, veins of calcedony trending N. & S. as well as in other courses, which could be often seen very distinctly for more than 100. yards, cutting not only across the hard level floor but through hills & valleys.

As we returned to camp Mr. Hayden and myself ascended the main bluff at several places in order to satisfy ourselves that I was not mistaken about the occurrence of bones and turtles in the highest bed. Upon examining the upper beds carefully we found bones and turtles (broken) not only in the upper bed but in several of the intermediate ones, and I have not much doubt but a careful examination would prove that they exist to some extent from the Palaeotherium bed to the highest bed in my section...

As we find it impossible to induce our men to remain any longer, and we have (including our Palaeotherium bones and turtles) as many specimens as we can haul to the Fort in our three carts, we have concluded to start as soon as we can pack up and get ready.

Sunday, July 10th.

Spent the day getting things ready and in bringing in from the field all our remaining specimens. Hope to be off early in the morning.

Monday, July 11th.

Rose at 4 oclk expecting to get an early start but our horses had strayed off and it was about 7. oclk A. M. before we could get everything ready to start. Upon putting all our specimens into the carts, found them very heavily loaded. Made an effort to ascend the hill out of the valley of Bear Creek but stalled before we had ascended half way up with the first cart — although there were four horses hitched to it. Took out and carried up a part of the things and made another effort to go up but failed. While we were trying at this cart one of the others which we had left standing on the hill side some distance below us, with one horse still in the shafts, commenced backing down the hill and turned over.

We now took four horses, and after righting up the lower cart tried to haul it up but could not induce our horses to move a peg above the foot of the hill. Seeing it was useless to attempt to ascend this or any of the other steep hills between here and the Fort so heavily loaded with such balky horses and unskillful drivers, concluded to open our packs of Palaeotherium bones which we considered less valuable than others, select out the best, and throw the others away. After thus lightening our load of about 100 lb. weight we took about half of the remaining load out one of the carts and by dint of a great deal of whipping & cutting, with no small amount of oaths in English, French and Sioux by the men, succeeded in reaching the summit (with 4. horses), after which returned with the empty cart after the remainder of the load, which was in the same manner hauled up. By thus making 6. loads of the 3. we at last had the satisfaction to see our things all on the summit. It was now 11. oclk A. M. and we had not yet eat[en] any breakfast. We had had plenty of time to do so, but we loaded up with the intention of driving on to Sage Creek to breakfast and as we did not know at what moment the men would come back with the horses did not eat. Drove on over a comparatively level prairie (with a few exceptions in the way of hills and valleys) where we had some tuging [*sic*] to Sage Creek where we arrived at 2. oclk P. M. Very tired and hungry. Mr. Hayden & myself have to walk the way to Ft. P. as my riding poney [*sic*] had to be put into the gears to supply the place of the horse we lost at Bear Creek. About this, however, I care but little if we can only reach the fort with our present load of specimens in safety. Although we have as much as we can haul we could have made a much more valuable collection, consisting of an entire load for our 3 carts of good heads instead of making up not less than two thirds of our load of large Palaeotherium bones, if we could only have induced our men to remain a few weeks longer while we sent in loads to the Fort. They were however afraid of the Indians, and would not consent to stay any longer for any compensation.

Had fine view of Bad Lands to S. E. of us on head branches Sage Creek as came along over high prairie between Bear and Sage Creeks. Blk. Hills seen off to SW. Although it was almost insupportably hot on the prairies where we were traveling could see snow on the Blk. Hills.

Camped on R. banks of one of the forks [of] Sage [Creek] on beautiful grassy spot beneath spreading branch of a few scattered cottonwoods of rather stunted growth. These cottonwoods, the only tree or shrub excepting the Buffalo berry... larger [than] the low bunches of shrubby sage... to be seen, on any of these streams, and even these are very rarely met with. Our horses having been nearly worked [?] out packing in specimens from the Bad Lands at Bear Creek, we will probably have to remain here tomorrow to allow them some rest.

Soils developed on Pierre Shale of Cretaceous age buried by the "Paleotherium Beds," of Oligocene age. The latter can be recognized by their light gray color and smooth, rounded slopes. The layered rocks in the distance are younger Oligocene in age. Refer to the Geologic Time Scale, Appendix 2. (Photograph by Richard C. Anderson)

Tuesday, July 12th.

Concluded to let our horses rest today. Toward evening when it was growing cooler loaded up with intention of going on about

3. m. to camp for the night where we would be beyond all the bad hills of Sage Creek, and from which place we could go on to Bull Creek before breakfast tomorrow. One of our young men (Isadore) had started out hunting on foot during the afternoon, and had not yet returned, but we expected to meet him before going far. We had barely started when we saw him coming down the hill covered with blood & loaded with meat. He had killed a fine fat Black Tailed Deer and was bringing in the hams, which were as much as he could carry, as it was a very large one. After driving on about 2. m. over some very bad road where I expected every moment some of our carts would be broken, we met our young Indian at the other branch of Sage Creek and were all very much disappointed to learn from him that some pools of good water where we had expected to camp up on the high prairie beyond the valley of Sage Creek were now dry, and as it was now too late to think of reaching Bull Creek tonight over such a rough hilly road there was no other alternative but camp another night at Sage Creek. This was by no means pleasing news, for the water of Sage Creek which was all we had to drink here, is so strongly impregnated with sulphate — magnesia? — that we were already purged by it until we were so weak we could scarcely walk, and besides it is exceedingly nauseating and unpleasant to drink. So is Bear Creek (and all the streams that head in Bad Lands), but there is a good fresh spring there.

Soon after stopping I noticed that a wheel of one of our carts, about which I had already felt some uneasiness, was about to break. Rondell however, who is a kind of universal genius, and always ready for any immergency [*sic*], hunted up some strong raw hide thongs, and bound it up so as to appear quite strong. If it only takes us out of Sage Creek, and across Bull Creek, it may last until we reach the Fort.

Many Helianthus in flower along bank of creek. One large species like the cultivated.

Wednesday, July 13th.

Arose at 3. oclk and were under way by 4. oclk. Ascended the last bad hill just as sun rose. Fine view of Bad Lands off to S. and of Blk Hills which bound the W. & S. Western horizon…

Arrived at Bull Creek (12. or 15. m.) by 9. oclk, where we stopped for breakfast after ascending the hill which we had so much dreaded. All hands were encouraged for we were now over the worst of our difficulties and had (comparatively) good water to drink.

I am now satisfied that the lower part [of the] section exposed here on Bull Creek is the same yellow & redish [*sic*] clays in which we found the large nodose Baculites (beneath the Tertiary) at Bear Creek, and that much of the high country over which we passed between Sage and Bull Creeks is immediately underlayed by some of the same beds of Tertiary, though there may be no fossils. Can sometimes see White Bluffs over the Cheyenne, which are doubtless Tertiary, though owing to the genl. dip being to S.E. I do not think these beds can be very thick there.

At about 2. oclk P. M. a company consisting of some 3. or 4. white men, as many half breeds together with some 3 or 4 Indians and quite a cavalcate [*sic*] of squaws and half breed children came along on their way through from Fort Laramie to Fort Pierre. They had about 50 head of horses, 2 wagons and 1. cart. One of the men was Isadore's brother. He has some five or six horses not in use and generously loaned Mr. Hayden & myself riding horses. They have been traveling very hard today and will not start from here until tomorrow. In as much as it is very desirable for us to travel with them in order to avail ourselves of their assistance in case of any accident to our horses or carts, we have concluded to wait for them. They bring the news that there was some fighting between the Siouxes and the soldiers at Ft. Laramie and that the Indians are manifesting a more hostile disposition than at any time for some years past. In the aforementioned fight 3. Indians were killed and their band has sworn vengeance against the Whites. A part of this band is now over amongst the Black Hills within a day's journey of us. We would rather not meet them.

These men report large companies of emigrants crossing the plains. Emigrants had stopped for a short time near Ft. Laramie in consequence of hostile disposition of Indians. Plains on other side Platte [are] black with buffalos.

Thursday, July 14th.

Started before sun rise and came on to the head of Little Missouri (18 m.) where we halted for breakfast. Rode off to right of the road about a mile, near Pinos spring (2 or 3 m. on other side) to examine some Bad Lands where the Tertiary beds are exposed. This was as I supposed the place where I had collected some small univalve shells in fragments of white limestone as we went out, and I was anxious to give the locality a more thorough examination in order to ascertain the exact position of this bed in relation to the other beds which I had studied in the Bad Lands on Bear Creek. Upon arriving on the spot I was disappointed to find it was not the same place, as I could find no fragments of limestone. Upon descending into the basin I soon discovered that the beds were very similar to many of those seen on Bear Creek above the Palaeotherium bed, I. E. A series of white indurated clay beds alternating with indurated clay of slightly more yellow tint (I. E. same col. as turtle bed) and others with slight greenish tinge. In some of these beds, especially the lighter colored, there were many concretions[10]... One bed in which I noticed these concretions was apparently composed of highly indurated clay having in places much sand in its composition. In this bed, which was not the lowest by 15 or 20 feet, I saw quite a number of Palaeotherium bones apparently exactly similar to those seen on Bear Creek. I also picked out of the clay 16. or 18. feet below the Palaeotherium bones, the under jaw of a small animal which did not appear identical with any found by us on Bear Creek. It is possible that it may have fallen from some of the higher beds but I am inclined to think it was in place. The Palaeotherium bones seemed to be abundant for I saw them at more than one place and Mr. Hayden saw them in the same positions some distance further round one of the buttes... I was anxious to examine these beds more carefully and to take detailed section. But the carts, about [which I] felt much solicitude whenever they were not under my own eye, were now a long distance ahead and I knew I could not more than overtake them before they would reach some bad places in the road where I was afraid they would upset. Consequently I picked up one of the vertebrae of the Pal[aeotherium] — and the broken jaw of the smaller animal and rode on after the carts.

At about 12. oclk we descended a very steep hill into the valley of the headwaters of the Titon or Little Missouri River. Pinos Spring is perhaps the extreme source of this little river. This spring is up on the high country and runs several miles before it passes down into the valley. After going down into the valley [it] went on some 1. or 2. miles to a pool of clear and cool water (in bed of a small stream now dry with exception of this and a few similar pools) about 7. feet deep and containing some small fish. Here we halted for breakfast and to let our horses rest and feed till evening.

After breakfast I walked to nearest hills or bluffs near the spot where I had taken a hasty section as we went out. I happened to find a much better exposure and noted down with much care the following section. (I do not now consider the sec. taken as [I] went out [to be] accurate.)

[Geological notes.]

In coming down hill to camp in evening saw something white, upon examination found it to be a human skull. Probably one of [the] Sioux nation.

Friday, July 15th.

Started by day light and came on to Grindstone Creek (where we arrived at 10. oclk A. M.) to breakfast. Out of fresh meat, for first time. Isadore started hunting, hope he may kill a good supply as game is growing more scarce as we approach the fort. A man belonging to Co. on the way to Fort killed an Antelope.

As [we] came by Grindstone Buttes, I rode a little out of the road to examine them again. There are 3 of these Buttes 300 yds. to ½ m. long ¼ as wide with a space of about ¾ of a mile between. Range E. 20 S. and W. 20 N. a few other conical hills off in range to N. W.

[Geological notes.]

Buttes flat on top. Had an indistinct view from top of Butte of some of the peaks of Bad Lands off to E. of South, far over and beyond range of bluffs on other side Little Missouri valley. Saw these as I came out but supposed them to be large trees...

Came to Big Cottonwood to camp all the way over undulating country underlayed by plastic clay.

Sat. July 16

Started before sun up and came to Ree Creek, to breakfast about 12 m. Ground exceedingly dry and cracked often from 1. foot to 18 inches deep. Soil, though dark, being nearly of the same col. as the underlaying plastic of which it is composed, does not appear to be rich or productive. This however is occasioned to some extent by the deficiency of rain. Great many Rudbeckia linta in flower. This is the celebrated black root so much talked of by the Traders and Indians as a remedy for the bite of the Rattle Snake. They administer it internally as well as to the wound, by chewing it and swallowing the juice and by applying it to the bite. I am inclined to believe there may be some virtue in it for I find by chewing it that it is a powerful and pungent stimulant.

Camped for the night at Bessi [?] Creek, having traveled about 26 m. which with our heavy loads and mean horses is doing very well. The party traveling with us killed two antelopes, one of which was a fine large buck with mature horns. Saved his head.

Sun. July 17.

Started at usual hour and came on to Frozen Man's Creek about 15. m. to breakfast. As it was rather too long a drive for our tired horses to go on to Willow Creek and there was only one place on the route where we could get water we concluded to only go to this point this evening although there was no wood and it was only 3. or 4. m. from where we took dinner. Isadore burned up his lodge poles to make a cup of tea for supper.

Mon. July 18th.

Started before day light and drove on to Willow Creek before breakfast. Morning cool and damp. There was a dense fog (the first we saw on the prairie) and a damp wind from the S. E.

While at breakfast the Prince of Nassau and party came along on their way through by Ft. Laramie to Great Salt Lake. After spending some time at the Salt Lake he intends going on through Mexico. His party consists of himself, Secretary and 3. men together with a half breed and interpreter.

Learned from the Prince that the companies boat had gone down only three days since, and that Dr. Shumard & Mr. Hendricks [*sic*] had gone down on it. Also that Dr. Evans was still waiting at the Ft. until the Blk. feet Sioux and some other unfriendly tribes now on their way to the fort come in. They are coming the same route he has to travel and he does not wish to meet them. These bands numbering 1000 warriors are now on their way to the Fort to chastise (as they say) the Two Kettles & No Bow and some other bands camping there, because they receive presents from the government and [are] unfriendly with the whites.

After breakfast came in to the Fort where we arrived at 11. oclk A.M. Found about 150 lodges of Indians consisting of Two Kettle, Yankton, Brule, & No Bow bands camped round the Ft. As we approached the Ft. our co. and those traveling with us drove up close together and hurried into the Ft. as fast as [we] could.

Towards sun set news came in that the unfriendly bands consisting of about 1000 to 1100 warriors are now camped at Willow Creek where we took breakfast. They will be in in the morning. Indians here very uneasy.

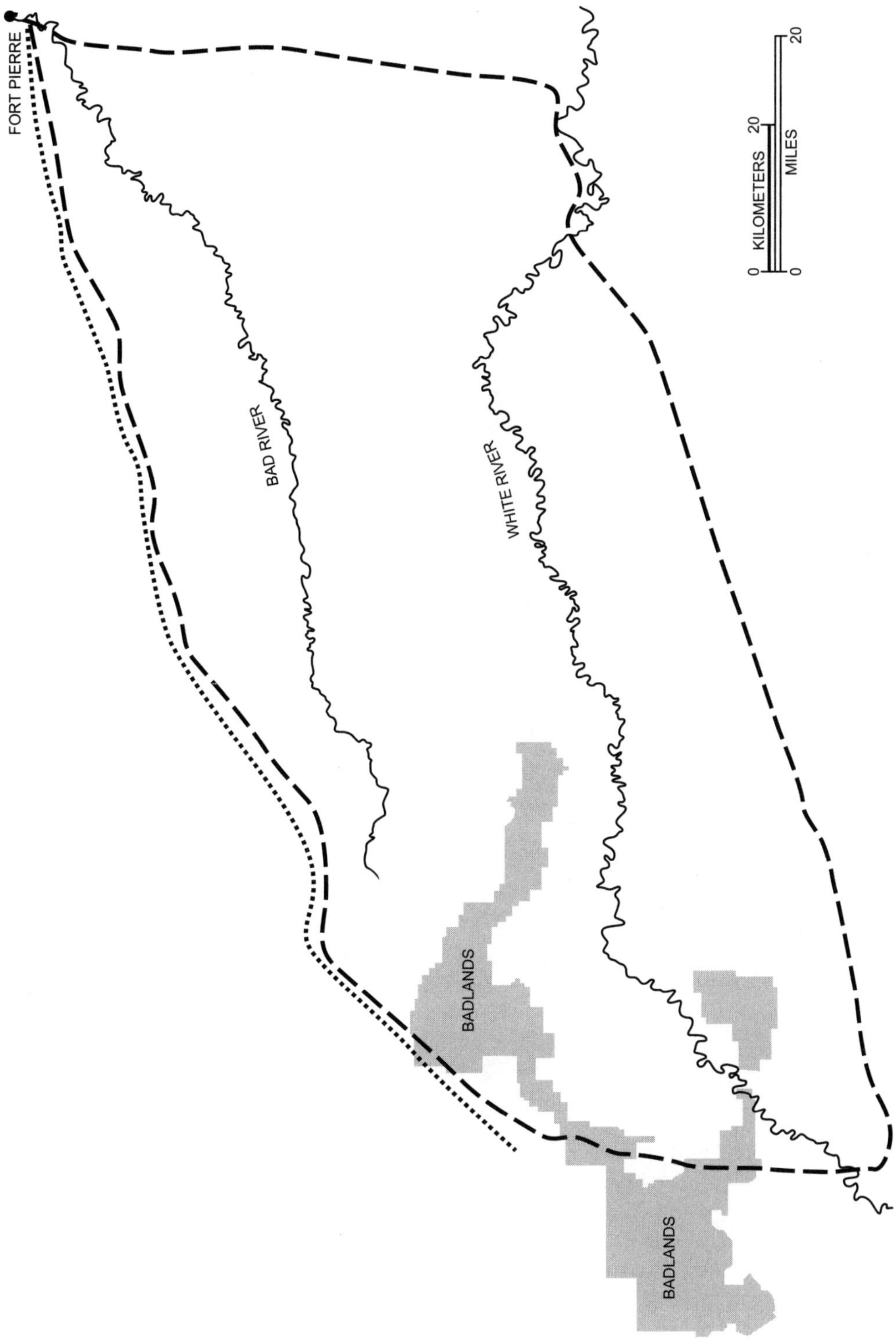

Routes of Meek and Hayden, 1853 (dots), and Hayden, 1854 (dashes) from Fort Pierre to the Badlands.

8

Downriver to St. Louis

With the entry for July 18th, the first part of Meek's 1853 diary ends, though on a later page of his notebook appears a "Memorandum of things left at Ft. Pierre."[1] Unfortunately his sketches of the Badlands have not been found.

The second part of Meek's 1853 diary begins with an entry dated July 27th. Thus there is a break of nine days in the record. Meek and Hayden probably spent all of this time at Fort Pierre. Since they had returned to the fort a few days too late to board the *Robert Campbell* on its return trip, they decided to transport their heavy collections downstream by means of a Mackinaw boat, a small craft built of sawed planks commonly employed for bringing cargoes down the Missouri.[2] The gap in the record from July 18th to July 27th leaves us in the dark with respect to their experiences at the fort, though much or all of their time must have been employed in preparations for the trip to St. Louis: buying provisions, hiring boatmen, repacking their specimens, and finally loading for the trip. The diary gives no information about the number of boatmen or the makeup of the crew, though it does comment favorably on their proficiency (e.g., on August 2nd Meek noted, "River very full of snags. Thought several times today we would certainly run upon some of them, but a skillful pilot and stout rowmen saved us").

This part of Meek's diary, like the first part, is devoted largely to scientific observations, but interspersed with these is enough information about the downriver trip to provide a satisfactory continuation of the narrative. And this information includes much of human interest.

Though Meek and Hayden were retracing their route of a few weeks earlier, they were doing so under very different circumstances. The Missouri was now at a low-water stage and they traveled with the current rather than against it. They were in a small boat, and they themselves had to make the decisions regarding when to travel as well as when and where to put ashore, quite unlike traveling on a comfortable steamer under a veteran pilot and captain and with a large crew of boatmen.

Having to grapple at close quarters with this mighty and capricious river made them more aware of its dangers. Hence Meek's repeated allusions to the innumerable snags in the bends, and to weather changes that at times threatened their venture. It was a trip that made them wiser in the lore of the great river.

From their little craft they looked out at the same alluvial islands, prairies, and bluffs as before, but now from different angles, providing new vistas that Meek deemed sufficiently important to describe.

Their stops "to dine" or to camp for the night were often made at locations where rock outcrops occurred, permitting study of the geological formations along with collecting.

Meek wrote for himself only, of course, making notes for his future reference and observations on subjects that particularly interested him. His hurried jottings reveal him as a practical man of science. His notes relate not only to his prime interests, paleontology and stratigraphy, but also to the soil and its fertility; to the mineral resources, especially coal; to the topography and climate; to the native vegetation and the crops observed at those scattered spots where cultivation was being attempted by Indians, whites, and "half-breeds." In conversations with "some intelligent Mormon missionaries" bound for England and France he discusses the numbers and migrations not of the Saints but of frogs. He notes that Independence, Missouri is "a very good locality for Carboniferous fossils" and that Jefferson City is "underlayed by heavy bedded white soft limestone." The loss of ground by the river towns through erosion by the river fascinates him. He is charmed by Cambridge, "one of the most desirable sites I have seen on the Missouri."

In line with the times, Meek noted the close relationship existing between "the fever" (malaria) and geographic conditions, without understanding the true nature or cause of the disease. Where the Missouri valley was wide, and its floodplains extensive, fever was rampant among the scant populace; where the floodplain was restricted or absent, as in the upper reaches and at a few places below Fort Pierre, conditions were "healthy." The fertile bottomland soil and the rank growth of vegetation on it were suspect; for the rest, the fever was a mystery — a plague to be accepted as inevitable under geographic conditions that were well-recognized.

The downriver trip, as recorded in Meek's diary, may be traced in the following paragraphs:

> Wednesday, July 27th [1853]
>
> Everything being on board of our Mackinaw boat, we started after breakfast, and came on down near the head of Red Cedar Island, about 35 miles below Ft. Pierre by water and about 24 m. by land. Country along both sides very broken and hilly, has barren aspect. Only few cottenwood [*sic*] and willow along shore, with many Bull berry (Sherperdia argenta) which are full of partly ripened red berries. Hills and country back as far as can be seen barren and destitute of trees. Valley generally not more than one and [one-]

Pages from Meek's 1853 diary, partially covering July 27-28, with geological sections. (Courtesy of State Historical Society of North Dakota)

half to three miles wide. About 9 miles below Ft. P. saw many boulders along hill sides.

A short distance above our camping place [we] stopped on left shore. Found hills composed of gray and bluish gray Cretaceous clays, same as seen all the way up to the fort. Exposures not very good here, many slides.[3] Upon ascending hill, found slope all the way up to summit and all the highest hills thickly strewed over with large granitic boulders.[4]... Along base of hill, near water's edge and for some distance up side of slope and sometimes in place [are] nodules and beds of harder rock apparently composed of same material as clays, with more iron, in which we saw many imperfect specimens of Inoceramus, Baculites, etc.... After eating dinner here we came on and stopped at a better exposure on the other side, where the following beds are exposed (a short distance below the first):

[Geological notes.]

About 15 Miles below Ft. Pierre passed the trading post of the opposition company, where the steam boat, the St. Ange, was discharging freight. When we were about to pass, some person on the boat hollowed to us and said there were some letters on board for some person on our boat. Upon landing and going on shore, we were pleased to receive news from our friends in the States.

Thursday, July 28

Started about daybreak and came on down to the Great Bend before breakfast. Stopped at the same high bluffs where we collected some fossils on our way up. Wishing to examine these beds after having studied the Cretaceous deposits higher up the river, we landed.[5] Found the following beds exposed:

[Geological notes.]

Friday, July 29th

Saw bluffs along right shore all along, 6 or 8 miles below bend, of same gray marl, which rise at some places as much as 50 feet above water, same drift beds on top, mentioned in above section. Country [farther] back broken and hilly, hills perhaps 100 to 200 feet high. On opposite side, terrace about as high as bluffs on this side, with higher country seen in the distance. Atmosphere yesterday and this morning filled with smoke. Prairie on fire some place near.

About 10 m. below camping place stopped for breakfast... While men were cooking breakfast, ascended the hill and found the following section exposed:

[Geological notes.]

Belts of trees becoming more frequent along margin of river. Valley about 3 m. wide... Passed mouth [of] White River at 1 ½ oclk.[6] (Raining.) Saw here the following section:

[Geological notes.]

The slides that have sometimes taken place in these beds might very easily lead one to think there were many disturbances and often a considerable dip[7] where nothing of the kind exists.

[Sketch of slides.]

Came on down to Bijou Hill where we camped for the night[8]... While men were cooking supper went out to examine Bijou Hill. Found it a very fatiguing undertaking to ascend to top. Distance from river to summit not less than 1¼ miles and I think the perpendicular elevation cannot be less than 1000 feet. Distance of this Hill, which is on the left shore, from mouth White River about 15 miles. Upon ascending hill had the following:

[Geological notes.]

I have no doubt, judging from the surface appearances along the slope, but there is at Bijou Hill an uninterrupted series of all the beds from the gray marl on up through the beds seen at Great Bend and at other places below, to the Tertiary whitish clays, which crown the summit of this and various other hills of nearly the same elevation on both sides of the river. I thought when I first saw thin white beds on the summit, from the river, that they were Tertiary. In this opinion I am confirmed by the teeth and bones found in them.

I have looked in vain, along in this part of the country, for the blackened hills [Burned Hills] mentioned by Nicollet.[9] It is true that the surface of many of these hills is nearly black when seen at a distance, in consequence not only of the natural col[or] of the clays but also from the great abundance of broken up fragments of argillaceous carbonate of iron, but there is no evidence that heat has had anything to do with this. It is true that there is occasionally (only at one or two places) between Scalp. Mt.[10] and Great Bend places where small local lignite beds have burned out and imparted, to beds immediately in contact, a kind of brick-red appearance, but I am convinced no such agency has had anything to do with the cause of the peculiar physical features of the country referred to by Nicollet.

Sat. July 30th

Breakfasted 10 miles below Bijou Hill...

Country for some distance above Scalp Mountain not so barren in appearance. There seems to be not so much naked black hills. The face of the country is quite as broken and hilly but more grass and vegetation... about 45 miles below mouth of White River stopped for dinner on right shore, where saw the following section:

[Geological notes.]

At 4 ½ oclk. passed Scalp Mountain. Upper part shows the same Tertiary beds seen on Bijou Hill. After passing this mountain the country becomes lower, but still broken and hilly. Same gray marl bluffs (fish scale bed) [seen at] base of hills along in this vicinity. Camped on right shore 6 m. below Scalp Mountain at what appears to be a drift terrace, about 30 feet high.[11] Scalp Mountain still on right and the highest elevation anywhere to be seen. The Sioux exterminated on the summit of this mountain an entire band of Rappahois who had established their village here about 50 years since.

Sun. July 31st

Morning cloudy, dense fog so could not see to run until about 8 oclk. Country lower than above Scalp Mountain. About 10 oclk. passed Towerhill about 18 m. below Scalp Mountain.[12] It is some two or 3 m. back from the river and appears to be a naked conical hill composed of yellowish clay. The shore being lined with snags and the water very swift, could not have an opportunity to stop and examine it. Took sketch. Took breakfast. In sight, some four or 5 miles below.

[Geological notes.]

Ten miles above the mouth of L'Eau Qui Court on left shore stopped at base of bluff where saw the following section:

[Geological notes.]

Bluffs presenting similar appearances to that where this section was observed [were seen] all along the shores either on one side or the other all day. Passed the mouth of L'Eau Qui Court[13] a little before sunset, and camp on same side at mouth Bazil Creek a short distance below, where we had exactly the same rocks mentioned in section 10 miles above...

The physical features of the country change somewhat as we pass down the river. The hills become lower and present more gentle slopes, whilst the valley grows sensibly wider so that there are occasionally some very pretty level bottom prairies. Timber is also more abundant along the river banks. Just opposite the mouth and below the mouth of L'Eau Qui Court, wide level prairie bottoms, scrubby oaks begin to make their appearance.

Mon. Aug. 1st

Started early and came on down 15 miles where we land on a low island covered with cottonwood trees, for breakfast. Left shore for some distance above [is] level prairie elevated some 30 feet above the river. This prairie appears to be underlayed by drift as nothing is seen along the bank but a bluff sand, gravel & clay — mostly sand and clay with a few pebbles — at water's edge some small boulders. On the other side the river, highest part of the country elevated perhaps 300 feet above the river, broken and hilly, gray & buff col[ored] marl bluff along shore. Some 13 m. below where we took breakfast there came up a southeast wind which made it rather dangerous to run, and we were compelled to lay up near a bluff [of] gray calcareous marl on right shore, in sight of Daurion Hill.[14] Not long after we stopped it commenced raining.

Collected here some oyster shells in the thin seams between layers of calc. marl... Fish scales all through both parts of the bed. No rocks to be seen in place above these steep marl bluffs, only grassy slope to the highest part of the country.

Came on down after shower and landed on Daurion Bluff on left at 3 oclk. Found 5 Indians, 3 squaws and several children of Yankton Sioux living here in an old abandoned trading house. All the balance of the band are up at Fort Pierre where they have gone to receive their presents. They had been there some days before we left. Those here had remained to work their corn of which they have quite a large crop (for Indians). They also have pumpkins, beans, etc. growing.

When we landed they all came down to the boat and we gave them some provisions, as they said they were out of food...

Camped for the night about 6 m. below Daurion Hill on left shore. Country as far back as can see on this side beautiful undulating prairie elevated 60 to 70 feet above river, and gradually rises as you go back. Soil very black & rich. About the first good farming land seen on our way down. On other side somewhat more hilly and elevated.

A short distance above our camp on same side there is a low bluff where the following section may be seen:

[Geological notes.]

Whilst walking along the shore at the base of the 30 feet bed, I picked up, amongst the loose pieces of marl from the beds above, a fossil bone about 2 inches long. It is an entire bone having both articulating extremities entire, is mineralized and quite heavy. Looks like one of the leg bones of large turtle, and I am inclined [to] think it was brought down and left there by the river from the Tertiary beds higher up — though it is possible it may have fallen out of some of the Cretaceous beds.

Musquitoes in swarms — find it impossible to write or be still a moment out of a musquitoe bar. Cloudy & warm — some distant thunder & a little rain during the night.

Tuesday, Aug. 2

Started by daylight. Passed the mouth [of] James River at 7. oclk.[15] Water of James R. clear. Country here on both sides beautiful undulating prairies with belts of timber along the river. Five miles below the mouth of James River, landed at the base of a bluff on right shore for breakfast. Whilst men were cooking I examined rocks exposed here, where I found the following section:

[Geological notes.]

Low bottoms & island covered with trees. A few miles below saw same beds of last section where the lower dark shaly bed is as much as 60 feet in thickness and has some very large calcareous concretions (3 feet long, 2 feet thick) towards the lower part. Some of these concretions have a soft incrustation of ferruginous matter spangled with numerous long slender crystals of selenite arranged and grouped so as to form beautiful stars or rosettes.

At about 2 oclock passed the mouth of Vermillion River.[16] Saw here encamped on a high prairie bluff elevated some 40 feet above the river, with beautiful grassy slope in front, some 8 or 9 lodges of Sioux. Wishing to avoid them we came down on the right hand side of an island which occurs here. Same range of bluffs where the last section was taken to be seen on right hand side a few miles above us and for a long dist[ance] below. River very full of snags — will think we are very fortunate if we reach the Council Bluffs in safety. Valley not less than ten miles wide, bottoms and islands low and clothed with trees.

About 5 m. by water below the mouth of Vermillion River and not more than 2. by land, high bluff on right, where the following section is exposed:

[Geological notes.]

Came on down and camped for the night on Elk Horn Point[17] on Huppan Kutey prairie on left. This is the largest bottom prairie yet seen on our descent. It is about 10 m. wide (back to bluffs which are low) and extends from the mouth of Vermillion down to the mouth of Big Sioux 30 m.[18] This prairie is almost perfectly level and only elevated about 10 feet above the present stage of water, and is sometimes (or at any rate a large portion of it) subject to inundation. The soil is exceedingly fertile, and at this time there is a luxuriant growth of grass & other vegetation as high as a man's head.

River very full of snags. Thought several times today we would certainly run upon some of them, but a skillful pilot and stout rowmen saved us.

Range of bluffs same as in last section seen all along the shore on our right. I have been looking all day for Dixon's Bluff described by Nicollet but have not been able to identify it. I have stopped at some three or four bluffs above but have not seen anywhere that the section exposed agrees with that given by him as occurring at Dixon's Bluff.[19]

Wednesday, Aug. 3^{d}

Rained this morning. Started after shower. River very crooked and every bend full of dangerous snags. Ran over one snag and against a rock in trying to land during a shower, but did no damage. Passed two or 3 bluffs along on right shore which appeared to be composed of same bed noted in last section but not near so high above water. The last bluff appeared to have some harder and thinner strata below, and I was very anxious to stop and examine them, but the current was very rapid and the shore lined with dangerous snags for a long distance. I would, however, have landed and walked up some distance to see it, but I was under the impression that the same line of bluffs were to be seen below the next bend. But I was very much disappointed to see nothing but a series of great bends with low islands and wide bottoms on down to the mouth of the Big Sioux where we arrived about an hour before sunset. All along our journey today the valley is very wide, country [has] rich soil, and a good deal of timber along the margin of the river and a great proportion of the level rich prairie.

We stopped a few minutes at the mouth of the Big Sioux where the following section is to be seen:

[Geological notes.]

I do not think it is more than 15 miles in a straight direction from the last bluff of Cretaceous strata seen by us to the mouth of the river, but owing to the numerous bends in the river it is perhaps not less than twice that distance by water. Saw a few Canadian Frenchmen living with their squaw wives here.

Came on down to Sergeant's Bluffs (5 m.) to camp.[20] These bluffs present a bold precipitous front not less than 100 feet in height, and are composed of the same yellowish siliceous marl seen above the sandstone at mouth of the

Big Sioux. Saw in it a few specimens of recent species of Helix. Upon the summit of the highest conical hill a single cedar post marks the spot where Lewis & Clark buried one of their men (Sergeant Floyd). From which circumstance the Bluffs here will perhaps be always known as Sergeant's Bluffs.

Saw here a gentleman who has contract to survey a portion of the public lands along on the Missouri and the Big Sioux. Also an American settler recently arrived and several Canadian Frenchmen who live by a kind of smuggling trade of whiskey with the Indians. Also saw a man (an American) by the name of Thompson, who is the first white man that lived here. He is a desperate character who has made a living by smuggling liquor into the Indian country. He murdered in the most brutal manner last fall a Mr. Norwood, Indian agent for the tribes upon the Upper Missouri, and still runs at large.

Walked about a mile down the river with our pilot and one of the hands to the cabin of a Mexican who is an old acquaintance of theirs. He was formerly in the employ of the Fur Co. at Ft. Pierre, but now lives here with his half-breed wife, a kind of half civilized life, cultivating a small piece of ground and carrying on a kind of contraband trade with the Indians. He is also a celebrated hunter. The soil being very rich here, his corn, potatoes, etc. etc. appear to be growing finely. He speaks French fluently, also Sioux and Spanish, and understands some English. Has plenty of horses, cows, pigs, etc. and appears to be as happy as a king. He was very much pleased to see his old friends and insisted upon having us all wait for supper which his half-breed wife soon had ready, and as none of us had tasted any vegetables or milk for a long time we made a very hearty meal. When we sat down to eat one of our comrades, an employee of the Fur Company who has been in the Indian country for a long time, remarked (with a tear glistening in his eye), "This table with its dishes, coffee pot, white cloth, etc. etc. recalls to my mind a thousand circumstances of my early youth, that occurred far away from here. This is the first time I have sit down to a meal upon a table for 3 years." Poor fellow, his food up in the Indian country, in the employ of the Fur Company, is hard jerked buffalo meat, and corn, and not always plenty of that.

We are now in the State of Iowa on left since passing the mouth of the Big Sioux. Country appears to be underlayed by the siliceous marl seen in the bluff here, and is rather hilly near the river but becomes more level [farther] back. Wide prairie bottom of very rich soil. Forest along the shore, very large black walnut and other trees indicating a good soil.

Thursday, August 4th

Came on through some very bad bends this morning where great skill was required in the pilot to dodge the numerous snags. Valley very wide. Bottoms and hillsides in sight of river apparently more frequently clothed with timber.

About 25 miles by land and perhaps twice as much by water below Sergeant's Bluff we came to some bluffs on the right hand side, where I went on shore and saw the following section:

[Geological notes.]

All below the lowest bed mentioned in this section [are] unknown, as all the lower beds are covered by a slide.

Most of the beds in this section are only local, for upon following the bluff some of them thin out in the course of 50 or 60 feet, whilst others increase in equal proportion. And at only a distance of two or 300 yards below, the same line of bluffs passes wholly into a soft heavy bedded yellowish sandstone, which breaks into [a] large column or masses 15 to 20 feet in height. I have never seen large heavy beds of rock wedge out to mere seams, or to nothing, so rapidly before, in any formation excepting the Drift. But this rock is evidently of more ancient date than Drift. As I saw no fossils in it I cannot form any ideas to what geological period it belongs. I regret very much that I had not time to examine these bluffs more carefully.

Camped for the night on right shore at Omaha Prairie about 30 m. by water and about the same by land above mouth [of] Little Sioux River. A short distance above where we camped there is a line of bluffs.

[Geological notes.]

River very crooked. Passed several bad bends today, one of which was so studded with snags that there scarcely appeared to be wide enough space between them to allow the boat to pass, and the current was so strong that we would have certainly been wrecked if we had touched any of them. The prairie where we are camping is quite level and large, bounded by a range of same bluffs seen just above us on the river, at a distance of 2 or 3 miles back (S) from the river. Immense bottom on other side of the river. Bluffs so far back as to be rarely seen on the other side.

Friday, Aug. 5th

Started soon after daylight but had to turn ashore and lay up soon, on account of fog. In a short time the fog cleared away and we started again, but after running about 20 miles were compelled to lay by on account of wind. While we were laying by about 10 m. above the mouth of Little Sioux River, two men who had deserted from the Fur Company at Ft. Pierre a few days before we left came by in a skiff. We hailed them but they would not land. In a short time the wind layed and we started, and after going a few miles we caught up with the two men in the skiff. At first they tried to outrow us, thinking we were some persons sent by the company to stop them. When we came near them they hailed us, and enquired who Mr. Hayden and myself were, and upon finding we did not belong to the company they came alongside. They said they were at Sergeant's Bluffs when we passed but had concealed themselves. They had nothing but a little flour to eat, and I gave them some meat.

Stopped for the night a short distance below the mouth of Little Sioux. Country seen today mostly low bottoms. Passed through one very bad bend full of snags. I notice that the right hand bends are nearly always free from snags, whilst the left hand bends are always full of them.

Saturday, Aug. 6th

Had quite a storm last night, chiefly wind (from S.W.) though there was some rain and thunder and lightning. We saw the storm coming up as we supposed from the west before we landed, and put in under the shelter of some low willows, but about dark the wind shifted round to S.E. so that we were exposed in such a manner that we feared the waves would sink the boat, and were kept awake nearly all night. Musquitoes very bad.

Started early and came on down about 25 m. by water to a point where there is a wide bottom prairie on right. (Nearly all the way above, low alluvial bottoms with falling banks, and numerous bends full of snags.) Just below the prairie, stopped to dine. Low gently sloping bluff and rather broken country near river on right. Low bottoms with dense forest on right side.

Valley seen this afternoon 8 to 10 miles wide. Hills only occasionally seen. Those on right more frequently [seen] than those on left. Right hand hills frequently clothed with timber. River very crooked, snags not quite so numerous. Bottoms sometimes extensive prairies and others covered with forest. [Have] seen no exposures excepting of loose sandy yellowish clay. At Old Council Bluffs where we passed a little before sunset saw on right hand shore a bluff about 100 feet high… On the opposite side of the river here on a low bottom there are a few log huts standing where there was once a Mormon village. As we passed up [the river] this spring, there were quite a number of California emigrants camping here, this being one of the crossing places on the route to California.

Sunday, Aug. 7th

Started very early and came on down to the Trading Post of Mr. Peter Sarpy, about 6 m. above the mouth of the Platt[e]. This being the end of our journey, in our Mackinaw boat, we had our specimens and other things taken on shore and stored until we can take passage to St. Louis on board of a steamboat which is hourly expected.

[Geological notes.]

Found here at Mr. Sarpy's, the Indian agent, Mr. Gatewood, very sick with fever. Hear of other cases of sickness here amongst the few settlers just coming into this neighborhood. This may easily be accounted for when we take into consideration that the valley is constantly growing wider as we descend the river from the mouth of Vermillion River, and that the soil on the numerous low alluvial bottoms as well as upon the higher part of the country is very rich and supports a rank growth of grass and other vegetation. Above the mouth of Vermillion where there are no bottoms of any extent and where the whole country is barren and dry from small quantity of rain, sickness is almost unknown. Although it is more sickly here at Mr. Sarpy's place than higher up, I do not think it is more so than in other new country of equal fertility.

There are some fine springs along the base of the bluffs here which appear to come from the beds of limestone. One that issues from the base of the hill about a mile below Mr. Sarpy's house appears, judging from the taste, to be impregnated with carbonate of iron. There is also some kind of gas constantly bubbling up in it (perhaps carbonic). Saw no feruginous deposit near it, though the shells of small species of Physa collected in it are coated with a precipitate of something like carbonate of lime. Water cool and not unpleasant tasted.

Corn, pumpkins, water mellons, potatoes, and other crops look very fine.

Wednesday, Aug. 10th

No steamboat yet. Went over to Ste. Marys with some gentlemen to examine the bluffs back of the extensive prairie bottom upon which the town is situated. The main object of excursion [was] to investigate some localities where coal is supposed to occur. In some low damp depressions and sloughs (extensive slough or pond back near bluff)...grass and other vegetation 8 to 12 feet in height. Weather very warm and sultry. Saw exposure of Carboniferous rocks along base of hills, perhaps as much as 12 feet above the level of the river.

[Geological notes.]

Mon., 15th Aug.

Weather very warm. Much sickness amongst Indians and whites connected with Mission & Agency here.

S. B. Robert Campbell made her appearance this morning — passed on up to Council Bluff City landing.

Tuesday, Aug. 16th

Boat returned at 3 oclk. Took all our things on board and started. Noticed just below the mouth of the Platt[e] River same beds of sandstone, shaly matter, etc. etc., with pink clay beneath, which had perplexed me so much above this [place]. I am inclined to regard these beds as Carboniferous, though I have not had a sufficient opportunity to examine them.

Passed the mouth of Nishnebotna [River] about dark. Wide alluvial bottoms, with caving in banks. River crooked and all left hand bends full of snags. Boat runs over many of the small limber ones without doing any damage. River only approaches the bluffs on the right hand.

Wed., Aug. 17th

Rocky banks along right shore, being ledges 10 to 50 feet of attenuations of the beds [of] limestone and clay and shale like those seen at Ste. Mary. Rocky bluff on left shore — an unusual thing.

Had conversation with some intelligent Mormon missionaries on their way to England, France, etc. Was informed by them that they met millions of frogs several hundred miles west of the Missouri migrating to the west. It is a curious fact that I have heretofore neglected to note, that we have seen on our way down, all the way from Ft. Pierre to Council Bluffs, migrations [?] of these frogs along the shores. Was informed by Mr. Sarpy and others that they have been crossing the river since about the 1st of July — that they increased in numbers until about the last of July, since which they have been diminishing in numbers. Was informed by Mr. Curtis, Chief Engineer of the Pacific Railroad, that he has seen great numbers of them during his survey. They were often so numerous that it was almost impossible to step without killing hundreds of them. In damp places he estimated them at not less than 50 on each sqr. foot of the whole surface of the country. Springs were filled so that in order to dip water in the mornings it was necessary to take out sometimes as many as three gallons of them. Mr. Curtis says he heard of some instances where they had collected in wells to such an extent that it was necessary [to] remove them by buckets full every day.

Arrived at St. Joseph at 12 oclk. Town improving very rapidly, but is fast caving into the river. Was informed that the river has encroached upon the bank not less than 100 yds. since spring. Population 4,000.

At Weston at dark. Not quite so large a place as St. J. but as much business done. Built upon rocky foundation — Carboniferous. To [*sic*] dark to examine the rocks much. Very sickly. Stayed here all night.

Thur., Aug. 18th

At Parkville about breakfast time. Pretty place, on rocky foundation. Rocks Carboniferous... River bottom or valley from near Weston becomes much narrower, more timber, not so many snags. [Steamboat] ran against snags & rocky bank this morning & broke guard.

Passed mouth [of] Kansas [River] at 10 oclk. this morning. Stopped at Kansas a short distance below starting point to Santa Fe & California...

At Independence landing at 12 oclk... Rocks full of fossils, a very good locality for Carboniferous fossils...

[Reached] Lexington at sunset. Bed coal about 60 feet above water level...

Friday, Aug. 19th

At Miami at sunrise. Saw here from waters edge up as high as 15 to 20 feet beds of compact limestone, with seams and beds of chert of various thickness intercalated. Many crinoid joints. Heard of coal being found in the country back. Also along the bluff near the river.

[Reached] Cambridge at 8 oclk this morning. It is a neat little village on the right hand shore, on beautiful slope which rises back to perhaps 100 to 150 feet above the river. Houses white and neat, scattered about amongst groves of pretty trees. Large shipping warehouses. Country back rich, producing hemp, tobacco, corn, wheat, etc. One of the most desirable sites I have seen on the Missouri. Banks not so much inclined to cave in, valley narrow, bottoms not so full of sloughs, and healthy after coming below Weston.

12 m. below Cambridge passed Glasgow on left. A very pretty and business-like place. Situated upon a rocky elevation, or rather upon a terrace 40 to 50 feet above the river, with ledges of rock along the bluff. Seams of coal amongst the limestones.

15 m. below Glasgow, Arrow rock, at village situated upon a rocky bluff about 75 feet above riv[er]. Country back elevated from 75 to 150 feet above riv[er]. Soil very good. Rich country. Bluff shows 30 to 40 feet of thick beds [of] compact blue limestone, intercalated with seams and beds [of] chert. Healthy.

[At] 2 oclk P. M. passed Rockport on left. Perpendicular bluffs of horizontally stratified limestone, 75 to 100 [feet] in height, above & below the town.

[Reached] Jefferson City a little before sunset. Elevation 80 to 100 feet. Underlayed by heavy bedded white soft limestone.

[Reached] mouth [of] Osage River about dark.

Sunday, Aug. 21st

Arrived at St. Louis this morning at 11 oclk.

Thus the collections reached St. Louis safely, after having been transported nearly two thousand miles by cart, Mackinaw boat, and steamboat.

At St. Louis, Meek and Hayden each received a letter from James Hall, who had written them in the supposition that they would remain in the West much longer than they actually did. Hall's letter to Meek read, in part, as follows:

Olean, Aug. 12th, 1853.

My dear Sir:

I have received yours of the 7th and 19th June, and have provided for meeting the draft of $212.26 which you mention…

I am glad to see that your affairs are apparently working more prosperously, at least it would appear that Dr. Evans and his party are disposed to be on good terms with you. I learned one interesting fact that after Dr. Leidy had written to me that Baird obtained of the Academy and of Leidy $600 for the expedition of Dr. Evans. You may recollect that Dr. L. wrote me he thought he would raise a few hundred dollars to send another person with you. From your account of your outfit and men everything appears promising and I hope you may have all possible success. This will probably reach you to [*sic*] late to be of service to you while there but I hope you will gather facts to give some idea of the geology of the region and its connexion with the geology farther east. It appears that you were able to trace the Carb. limestone for a long distance; farther up the Missouri than I have before known of it.

One thing surprised me in conversation with Baird and this was that he did not understand that Shumard was [to] return by the boat. This you had written was their plan. With the money from the Government, the Smithsonian, & the Phil. Acad. Sc. they are prepared no doubt for a longer stay and I am expecting to hear that such is their plan…

I am still at some loss what to say to Mr. Hayden. If I knew that all my projects would turn out I would not hesitate to give him encouragement to remain. Dr. Kirtland is not willing to do anything and leaves the matter upon me. For my own part I think I could profit more by employing him in this region... And [if your] collection [from] the Mauvaises Terres is good and you think it not to be greatly enhanced by his remaining then you had better both return. Since the expedition has been undertaken I do not want it to prove greatly inferior in results to the others though there has been such a flourish of trumpets and resources besides for cash. If one or two hundred dollars could greatly enhance the value and importance of the collection I could not hesitate, but to go to much greater expense I do not feel to be right for me. You will judge of all this and be better able to decide than I can and on your judgment I shall rely and feel satisfied.

I think a full series of the Carboniferous fossils of Missouri would be of more value to me than the Cretaceous and others from Nebraska. I would like our Nebraska collection to be a good one, and if you and Mr. Hayden will prepare materials for a paper on the general geology, a list of the fossils, scenery, sections, etc. etc. I will help you to put [them] in such a shape as to be published.

I shall never work in that field and therefore have not the anxiety for all the facts, specimens, and detail that I have for those of the Paleozoic formations...

I will send forward to St. Louis a draft for you about the first of October so that it may be in time. I will probably send this to the care of Dr. Engelmann...

I am very sincerely your friend,

James Hall[21]

The letter to Hayden was written on the same day as that to Meek. It read:

Olean, N. Y., Aug. 12th, 1853.

My dear Sir:

I have been debating with myself a long time as to what I should say in reference to your remaining over winter in the Badlands. Perhaps before you see this you will have decided that question yourself however. Your very main object was to make a good collection of the vertebrate fossils of that region. I do not expect to work up this matter in detail and therefore I must, with my limited means, content myself with less than I otherwise would. If in the judgment of Mr. Meek & yourself the value of the collection would be greatly enhanced by your remaining over winter I would not much hesitate. Of this you being on the ground are much better able to judge than I am. I should hope that your summer work would present a very fine collection of these remains, and for other objects there is little encouragement. Dr. Kirtland will give no encouragement that he will do anything — indeed he says he has as much on hand as his means will permit. I have not the hundredth part of the means he has and I am compelled to work up carefully the details of Paleozoic [paleontology]. Therefore you see my position.

If you return and finish your Medical Studies this winter I think I can find something for you to do in the spring and if I am successful in my plans I may find you enough for a year or two in this line.

I shall do all that I can do I will assure you, but I am not certain beforehand of what I may be able to do. The summer will have given you an opportunity of seeing something in geology and you will be better prepared to work alone in some other field than when you left last spring. You will yet require some months to get acquainted with the Palaeozoic rocks before you can collect to the full advantage from these.

I feel that all the money I can spare ought to be devoted to these rocks and to the collection of their fossils. The field yet almost untouched remains one that I can do more good in this way than in expanding my efforts. Still as I before said Mr. Meek & yourself can judge better what is to be done or what ought to be done. I anticipate that the whole expense of this expedition will be nearly or quite twice what I anticipated when I began. Therefore this is some reason for desisting in the autumn if we have a good collection.

I can well sympathyse [*sic*] with you for I know how eager I have been to explore the whole world and to bring it together in my cabinet. I find however that is only a small part of this that one can do in a single life, and without means.

Dr. Newberry feels much interest in your success. He has been very sick during the summer but is now better. We have had a very pleasant meeting at Cleveland from which I am now returning and have been detained here some days by sickness. I shall be very glad to have a letter from you. There is no reason for letting Mr. Meek do all the correspondence, or at least why you should not give me as much more.

I am very truly your friend,

James Hall[22]

These letters from James Hall, and perhaps other letters they found waiting for them at St. Louis, brought them into touch with the world of science again, as also did their calls on Dr. George Engelmann, who came to their financial aid by advancing them funds on behalf of Hall. Thanks to Engelmann's assistance, they stayed in St. Louis only long enough to transfer their belongings and load of fossils to another steamboat that would carry them toward their destination, Albany.

9

Aftermath of the Journey

A few notations in Meek's diary sketch the more important facts about the trip of the two young men from St. Louis to Albany. At "5 ¼ P. M." on Tuesday, August 23rd, they started for "Cincie" (Cincinnati) on the steamboat *U. S. Mail.* That evening, just before dark, the steamboat grounded on Cumberland Bar. On the 25th Meek noted, "Got off the Bar this evening at about 2 oclk and passed Smithland." His next entry reads, "Arrived at Albany on morning 3 Sept. at about 2 oclk. Our collection arrived in safety some 2 or three days after."

On September 30th, Meek resumed more detailed entries in his diary under the general heading, "Memoranda of Weather, etc." Meek recorded daily observations of meteorological conditions at Albany: temperatures and precipitation, ground conditions, wind directions and velocities, clouds and cloud cover, and the occasional thunderstorms and displays of the aurora. His notes, generally covering the period from sunrise to midnight or nearly so, are a measure of his long working days, devoted mostly to drawing fossils but sometimes to preparing specimens and writing. The evenings were almost always spent in making casts of fossils.

In his diary Meek recorded things he learned from his reading, also his candid appraisal of recent publications. For example, for Oct. 28, 1853 he wrote, "I have just seen Marcou's geological map of U. S. etc. and accompanying explanations. I think it is a very poor production."[1] And for Nov. 4th he noted, "Saw Prof. Hitchcock's Geological Map [of] U. S. and of the world. I do not think much of them." Likewise in his diary he reported the visits of scientists and others, who called on Professor Hall or himself, or came to view Hall's "cabinet" of specimens. For Nov. 16th he wrote, "We were visited today by the Pope's Nuncio, The Bishop of Albany, together with many of the Catholic clergy, the Secretary of State, etc. etc. Spent the day after our visitors left the cabinet in studying the anatomy of Crinoidea. Made casts till late." On January 14, 1854, he noted, "Prof. Agassiz arrived. He and Judge Parker and the Judge's Lady came over to the cabinet after supper. All much interested in our collection from the Bad Lands."

Living for his scientific work and absorbed in it throughout his waking hours, this lonely man was seldom conscious of any other happenings (excepting always the changes in weather conditions, to which he was remarkably sensitive); though Sunday was a day of rest, he generally spent it "in the house," reading technical works. For November 25th he noted, "Made drawings of Crinoidea today. Casts at night. Forgot all about this being Thanksgiving until evening." For December 25th his entry reads, "I spent this Christmas day in the house reading W. F. Van Amringe's work upon the Nat. History of the Human race. I am satisfied from reading his writings as well as those of Smith, Agassiz and Morton on this subject and from my own observations, that at least four of the great subdivisions of the human family spring from as many originally distinct races and that no climatic or other local causes could ever have produced the marked differences now to be seen between the races." For Saturday, Dec. 31, 1853 [absent-mindedly recorded as Oct. 31] he wrote, "Spent the day drawing figures to illustrate the anatomy of Crinoidea. This is the last day of the week; the last day of the month and the last day of the year. This year has been on the whole to me of more than usual interest. I have traveled during the year about 8000 miles, seen much of the world, and learned some lessons in human nature as well as Science — passed some pleasant days and some very unpleasant ones; and I think I am altogether a wiser, if not a better man, in consequence of my experience during the past 12. months. What is to be my experience during the coming year, or whether or not I am to live to see its last day, is only known [to] the Ruler of the Universe."

Meek's diary makes mention of many days during the closing months of 1853 spent at work on the specimens collected the past summer. For example, on Nov. 30th he wrote, "While labeling a lot of specimens of jaws etc. identified by Prof. Leidy as Rhinoceras nebrascensis I discovered amongst them a part of a [specimen] having the characters of Archeotherium which must have been overlooked by Dr. L. If it is not a very young individual of A. Mortoni it must be an undescribed species. I am inclined to think it was not a young animal, but a new species which could not have been larger than a hare." On Dec. 7th he noted, "Spent the day mending broken specimens of jaws from Bad Lands. Made casts at night." On Dec. 28th he wrote, "The Crustacean found in the Cretaceous rocks at the Great Bend of the Missouri by Mr. Hayden belongs to the genus Pagurus." Most of Meek's diary entries, however, refer to the tedious work of removing matrix from the "heads" of vertebrates collected in the Mauvaises Terres.

A few letters also help one to follow the thread of events after the return of Meek and Hayden. Exactly two weeks after getting back to Albany, Meek wrote the following letter to Dr. George Engelmann:

Albany, Sept. 17th, 1853.

My dear Sir:

Inclosed you will please find a draft for the am't you had the kindness to loan me as we came on. Prof. Hall started yesterday morning to attend some business in Western Virginia, and requested me to say to you that he feels under many obligations to you for this and the various other favours you did us while at your place.

Our collections arrived here in safety soon after our return and Prof. Hall is very well pleased with it, though he regrets that we could not have stayed longer and explored more of the country. Prof. Baird was here yesterday soon after Prof. Hall left, to see our collection; he pronounced it magnificent, and vastly superior to all the collections heretofore brought from that region (He has not yet seen Dr. Evans' late collection).

We have better specimens of everything described by Leidy than anything figured by him, besides some undescribed things of much interest. We expect Prof. Agassis [*sic*] here as soon as Prof. Hall comes back.

Mr. Hayden is well and sends his respects. Prof. Hall says he will send you a suite of our plants, as soon as Mr. Hayden can have time to look them over.

In great haste I remain with much respect etc.

Sincerely yours,

F. B. Meek[2]

George Engelmann wrote in reply:

St. Louis, Sept. 30, 1853.

My dear Sir:

A few days ago I received yours of Sept. 17th with the enclosed draft. I herewith return you the due bill.

I am delighted to learn that your collection has turned out so well. How unfortunate that you could not stay any longer there!

Gov. Stevens' Party do not appear to [be] very harmonious; quite a number of his companions have left him, I learn, before reaching Fort Union; one of them, an artist, is now with Col. Frémont on his western expedition. The Col. had to return on account of ill health, but is better and will leave here next week to rejoin his party on the plains.

Give my respects to Prof. Hall. I shall be much pleased to receive a set of Mr. Hayden's plants…

I remain, very truly yours Mr. Meek,

George Engelmann[3]

Joseph Leidy was not slow in inquiring about the vertebrates that Meek and Hayden had added to Hall's collection, as is evident from the following letter:

Phil. Oct. 1, 1853.

Prof. James A. Hall,

My dear Sir,

About a month since I received for examination 10 large boxes of fossils collected by Dr. John Evans in the Mauvaises Terres. Besides numerous specimens of species already described, the collection contained several new genera and species which I characterized in the last number of the Proceedings of the Academy.

Prof. Baird has informed me you also have received your collection and if as you informed me you do not intend investigating them I would very much like to do it. If you give me permission to select out of the collection what is new to describe and figure, I will visit you at the earliest opportunity, for fear we may be anticipated in any discoveries through the collections sent to Germany. It would also afford me pleasure to label all the species which have been described.

Hoping to hear from you at an early period, I subscribe myself with respect,

Joseph Leidy[4]

Doubtless it was as a result of this letter that Meek could note in his diary on October 24th, "Packed up box of specimens from the Mauvaises Terres to send to Dr. Leidy." Thus a first consignment of fossils was made available for investigation by Leidy, presumably at the Academy of Natural Sciences of Philadelphia. How many other shipments followed this one is not known, nor has information been found concerning the financial transactions involved. Hall did not himself describe any of the vertebrates, and eventually he, no doubt, disposed of the greater part of the collection by sale or trade.

That the collection proved to be an excellent one is clear from Meek's letter to Engelmann and other statements. Hall himself, in writing to Hayden three years later acknowledged that the collection made in 1853 was "extremely creditable to Mr. Meek and yourself, much exceeding in value the one made by Dr. Evans."[5] Hall's biographer, John M. Clarke, termed it a "harvest," adding that "Hall had for his material share a large collection of unusual fossils with which to reimburse his pocketbook."[6]

The invertebrate fossils that Meek and Hayden collected on their western tour of 1853 were retained by Hall himself, except for duplicates that were doubtless disposed of in one way or another. These fossils provided the basis for a joint study by Hall and Meek, carried out in 1854 and 1855. The completed memoir was published at Boston in 1856 by the American Academy of Arts and Sciences.[7] The paper included a brief vertical section of the Cretaceous beds, which is now of historical interest as being the first section ever made in the Badlands region. As such it was reprinted many years later (1924) by George Merrill in his *The First Hundred Years of American Geology*.[8] Hayden had no direct part in this project for a number of reasons, among them the fact that while the paper was being written he was engaged in medical studies or was absent in the West, and also because Hall regarded him as a paid assistant rather than a colleague.

The memoir on the Cretaceous invertebrates was the only paper that Hall and Meek coauthored, although both men continued to publish voluminously. Most of Meek's work appeared under joint authorship with others. This is indicative of the deterioration in personal relations between Hall and his assistants after the 1853 expedition (and at least in part because of it). The difficulties between Hall and Meek culminated, at length, in a complete breach that was never repaired. The full story of this unfortunate development has not been told and is outside the scope of this study. Clarke gives a brief account of it in his biography of James Hall,[9] and some mention of it appears later in this book.

Hayden's disillusionment with Hall is likewise a recurrent theme for several years after 1853. It was particularly evidenced by his turning more and more for counsel and assistance to other scientists, notably Baird, Leidy, Meek, and Engelmann. This course eventually brought Hall's wrath down upon Hayden, with charges of gross ingratitude.[10] Thereafter, for a period, there seems to have been little communication between the two. However, Hayden, although he was excitable and quick-tempered, was not one to harbor resentment and rarely stayed at outs with anyone for very long. Furthermore, his manifold and expanding professional activities occupied him too completely to leave room for the nurturing of old grievances. (In this respect he differed from Meek, who, as his health failed and he became totally deaf, retreated into lonely isolation, and continually brooded over his differences with Hall.) In time, Hall and Hayden resumed professional contacts and correspondence, although they can scarcely be said to have become intimate friends. When, after Hayden's death, the Academy of Natural Sciences of Philadelphia established the Hayden Medal, it did not seem inappropriate to select James Hall as the first recipient (in 1890) of this award. Thus Hall's name heads the list of renowned scientists from many countries who have received the Hayden Medal in recognition of "the best publication, exploration, discovery or research in the science of geology and palaeontology."

It remains for us to trace Hayden, as best we can, through the months in 1853 after his return from the West. The long uncertainty as to whether or not Hayden should remain in the West over the winter was terminated by his decision to go East. How fully his plans had matured before he left St. Louis is not clear; Hall later (1856) asserted that it was by his advice that Hayden resumed medical studies, at Albany.[11] Probably Hayden's other scientific friends concurred in this recommendation, so patently in his best interest. Hayden himself, from the evidence of his many letters written in 1854, was quite aware of his need for a strong background in education and training, field experience, and communication with other naturalists, if he was to achieve his goals. But he was also impatient to get established in the work in which he had fairly revelled during his first taste of exploratory geology. His initial Western work had confirmed his earlier resolutions respecting a career.

Whatever the factors that influenced his thinking, Hayden carried out his plans with a decisiveness which might have surprised those friends of a few years earlier who felt "he seemed never to know what he was going to do with himself." His course may have been dictated to some extent by necessity, since he had no sure means of self-support in the West (whereas in the East there were friends and relatives from whom he could seek help), yet one can hardly doubt that by this time "the boy of the class" had become a resolute young man who could set creditable objectives, make firm decisions, and adhere to plans with dogged perseverance. These were traits which came to distinguish Hayden.

On September 16, 1853 (about three weeks after returning to Albany) Hayden wrote to Dr. Newberry requesting a certificate of private study that could be submitted to the Albany Medical College in partial fulfillment of requirements for the medical degree. Newberry replied from Cleveland, on September 21st, "I will cheerfully give you a certificate of study at my office during the years 1852 & 3, if that will serve your purpose though I don't see as that will make out your 3 years. You commenced in the spring of 1852 here, but had no vacations so you made out a year's study. In this state the same law exists, but, for graduates from colleges, an exception is made, regarding the college studies a part at least as preparatory to that of Medicine, and indeed some of them coming under that head properly."[12] Hayden must have addressed a request also to the Cleveland Medical College for certification of his first course of lectures, taken at that institution.

The fall term of the Albany Medical College opened on the first Tuesday in September. Hayden had returned to Albany just in time for it, and he remained in residence for the duration of the lectures, a period of sixteen weeks. The official register for the fall term listed "F. V. Hayden, A. M., Cleveland," as one of the 85 students. Dr. J. S. Newberry was named as Hayden's preceptor.[13] At the time, Albany Medical College constituted the Medical Department of the University of Albany. Founded in 1821 as a private institution, it had prospered and grown; in 1838 it had been chartered by the state of New York. Generously supported by the state and by local citizens, it was a superior medical school. Its excellent reputation was attested, for example, by the fact that Hayden's class included students from all parts of New York and from New England, Pennsylvania, Ohio, and states as remote as Georgia and Texas, and also from various provinces in Canada. For many years prior to the Civil War the number of students in attendance was well over one hundred. However, graduates per year averaged only about twenty-five.

The stability and high standing of the college reflected the devoted labors of Dr. Alden March, founder of the school and its president and professor of surgery, who from 1821 until his death in 1869 was "the driving force" of the college.[14] Dr. March surrounded himself with a faculty of distinguished professors, well trained in this country and abroad, who served the college loyally for many years. By contrast, many other medical colleges suffered badly from rapid turnover of staff. At the time of Hayden's attendance, the Albany faculty numbered thirteen (excluding Curators). Two of the staff were men of prominence in geology as well as in medicine: Dr. Ezra S. Carr and Dr. Ebenezer Emmons. Dr. Carr was Professor of Chemistry and Pharmacy; Dr. Emmons, Emeritus Professor of Natural History. Dr. Emmons, in particular, was a versatile scientist — a chemist of note, and the author of a *Manual of Mineralogy and Geology*. He had filled various chairs at the college, including, at one time, that of Professor of Obstetrics and Natural History.[15] Since these men shared Hayden's interests, he may have become well acquainted with them while at Albany.

Several unusual features lent distinction to Albany Medical College. The charter specified that all candidates for the medical degree should be examined orally by a staff of appointed "Curators" (physicians not on the faculty, generally five in number). This practice was continued for half a century. The institution sought to provide satisfactory clinical instruction (Dr. March and his associates have been credited with originating the college clinic idea). Also, the establishment of a hospital twelve years after founding of the college "was a progressive move that kept Albany in the van of American medical education. Without its clinic and hospital it might have justly suffered the fate of the backwoods schools. Instead, March and his associates met the challenge of their day and turned a country school into a city school with an extensive museum and library, satisfactory laboratories, and more than ordinary clinical facilities."[16] In 1873 the Medical Department of Albany University became the Medical Department of Union University.

These advantages and others were pointed out in the Catalogue and Circular for the Fall Term of 1853. It stated that the College was "pleasantly situated near the Capitol" and that its edifice was "spacious and commodious." Lecture rooms, laboratory, anatomical theatre and dissecting rooms, and the halls for Museum and Library were fitted up with all the requisites "to render them convenient and suitable for medical and surgical demonstrations and instruction." The Museum, "in the variety of its specimens, both natural and morbid, in the number of its delicate and minute preparations, and in the abundant means which it furnishes for ready illustration, is unsurpassed by any in the Union. The liberal endowment of the Institution by the state, has furnished abundant means for increasing the number and value of these specimens and preparations. Their selection has been made with great care, and, in many instances, under the direct inspection of the professors themselves, while abroad in the various countries of Europe."[17] The library in 1853 numbered nearly 5,000 volumes.

Facilities for dissection were under the supervision of the Professor of Anatomy and his assistants. An abundant supply of material for dissection was furnished "on the most reasonable terms." The Saturdays of every week were devoted to practical and clinical instruction, giving students the opportunity to witness a great variety of medical and surgical cases. It was stressed that "Persons who are really indigent, and who require advice or surgical operations are gratuitously attended to… The great advantages derived by the student from these cliniques are to be found in the intimate knowledge of the case and its history; the opportunity for witnessing a great variety of operations; the most approved method of dressing; and the medication proper to each case. From these clinical lectures and operations… the student derives an immense amount of varied practical knowledge."[18]

It was under highly favorable conditions, then, that Hayden completed the second of the two required courses of lectures. He also had to fulfill certain additional conditions required of each candidate, notably:

"He must deliver to the Registrar, at least six weeks before the end of the term, a thesis, written by himself, on some medical subject, and be prepared to defend it at his examination...

"He must pass a satisfactory examination on the several branches of medicine and surgery, before the Faculty and Curators of the College."[19]

A further stipulation, and one which in Hayden's case may have presented difficulty, called for payment of a graduation fee of $20 prior to admission for the examination.

All requirements were fulfilled by the end of 1853, and on January 8, 1854, Hayden could write to Spencer Baird, with relief and satisfaction, "I have now finished my course of Lectures and graduated very respectably, so that I am now free in that respect."[20] It is noteworthy that not a single letter by Hayden has come to light from the period he spent at Albany Medical College. Evidently he was so absorbed in the urgent task of completing the course that, during these months, he had to drop virtually all other activities. The only glimpses of Hayden from the concluding weeks of 1853 are provided by Meek's diary. Meek noted for December 23rd, "Mr. Hayden commenced working with the Collection of Plants from West." Again, on Christmas Eve he wrote, "Mr. Hayden at work upon plants." On Dec. 29th appears his final reference to this project, "Mr. Hayden still at work with plants."[21] Evidently Hayden, with lectures and examinations out of the way, spent his Christmas vacation in this manner in company with Meek. However, it is not known, from Meek's diary or any other source of information, if Hayden shared lodgings with Meek at the Office while he was engaged in medical studies; prior to the Christmas period, Meek made no mention whatever in his diary of the proximity of Hayden.

Hayden was fully committed to the pursuit of Natural History when he became a Doctor of Medicine, but his new degree was to signify more to him than merely a title and an open door to his chosen work. His medical training and introduction to hospital procedures proved extremely important to him in later years: during the Civil War when, as a Surgeon of Volunteers in the Union Army, he made an outstandingly successful hospital administrator; and during the many years he devoted to western exploration, when he was responsible for the care of his men and himself, particularly at times of accident or illness. However, as with a great many other naturalists of the period, the practice of medicine and surgery never became his major interest.

The friendship formed during the Badlands trip in 1853 between Meek and Hayden led to a fruitful association which endured until Meek's death in 1876. It "furnished an example of uninterrupted collaboration without a counterpart in the history of American Geology."[22] Of this unique partnership we will have much to say.

10

Call of the Wilderness

As has been noted, in January, 1854, Hayden found himself free: free in the sense that he had completed formal schooling and therefore could give himself unreservedly to natural history, provided, of course, he could also make a living.

Had he been content to settle down in some community to practice medicine, while pursuing natural history locally on a part-time basis, a common arrangement among doctors, probably he could have managed well enough. But this was not for Hayden, who, as one of his classmates observed, was "more content in the extreme than in the mean." Wrote J. Peter Lesley of him many years later, "There was a vehemence and a sort of wildness in his nature," and these terms well describe Hayden at the age of 24.[1] Though penniless, or rather, as he once put it, somewhat worse than penniless because in debt, nevertheless nothing would do but a return as speedily as possible to the work on which he had set his heart: explorations in the Far West. Toward this end he would entertain almost any kind of proposition.

The difficulty of carrying out such a resolution is hard to appreciate in our times, when any reasonable scientific project, however costly and ambitious, stands a good chance of receiving support from some source, a research foundation, federal or state agency, industrial organization, university, or professional society. Hayden could only hope to combine a career as explorer-scientist with eking out a living by finding a patron able and willing to assist him, by securing aid from the meager income available to the Smithsonian Institution or one of the few scientific societies (with none of which he was affiliated as yet), by appointment as surgeon on a government party of some kind, or through employment on a state survey. Positions of the kind he desired were relatively few and hard to get, at least by one of limited experience who as yet had no well-established professional reputation or record of publication. Hayden was realistic about these difficulties, but they served to stimulate his efforts rather than cause him to abandon hope. With the odds heavily against him, he sought to explore all possibilities in his search for an opening.

As a matter of fact, he had already anticipated this exigency the summer before when he tried to promote future explorations through the person of Major (or Colonel, as the case might be) Alfred A. Vaughan, Indian Agent to the Upper Missouri tribes, who had been a fellow passenger on the *Robert Campbell.* Major Vaughan, like Alexander Culbertson, had little scientific background himself but was genuinely curious about the regions into which his duties took him. Vaughan was sympathetic to Hayden's ideas, but their talks had not resulted in any written contract or other binding commitment. Now, in early 1854, when Hayden tried to reach the itinerant Vaughan by letter to confirm arrangements, he was unable to do so. In this uncertain state he investigated other possibilities but to no avail. As months passed and he still found himself "out of business," he resigned himself to the prospect of collecting plants for Professor Swallow of the Missouri Survey, only to reject this solution with alacrity when the missing Vaughan finally turned up and made a specific proposal much to his liking. Vaughan proposed that Hayden should return to Fort Pierre and make even more extensive explorations which would take him all the way to the headwaters of the Missouri.

Numerous letters, most of them written during the early months of 1854, tell of the hopes, discouragements, and frustrations which attended Hayden's efforts to get back into the West, and the eventual success which rewarded his persistence. Like the letters of a year earlier, when plans for the journey to the Badlands were taking shape, this correspondence is of general interest because of the references to contemporary personalities and events, especially men of science. Not least significant is the picture it gives of Spencer F. Baird, the patient, generous, and influential leader among American naturalists. Hayden's own letters reveal the tremendous impact which his first acquaintance with the West had made on him, his reaction to that "glorious country" which had awakened in him a desire to return so compelling that nothing could stop it.

Before Hayden had finished his medical course, Hall and Leidy exchanged views respecting his prospects for the coming season. Leidy wrote to Hall from Philadelphia, on December 19, 1853:

> I think it would be better for Mr. Hayden to examine some new locality than the Mauvaises Terres. With the small means which he might obtain he could do very little more than collect a few fragments of species already described. Mauvaises Terres may now rest until government sends out an Expedition especially to explore that territory, when with tools for quarrying, etc. and persons capable of selecting, they may make a rich and enormous collection.
>
> Positions for explorations frequently open, and the first of which I hear I will inform Mr. Hayden. Only last Saturday I was requested to recommend some scientific young physician to accompany Lieut. Strain in a Sloop of War to Panama, to undertake a Survey. But as the vessel was to leave tomorrow, Tuesday 20, I said the time was too short to permit me to inform any person...
>
> In regard to parties in Phil. subscribing for a Mauvaises Terres expedition, I think none can now be found; but if a new place was open for exploration I think 2 or 300 dolls. might be collected in aid of it...[2]

In January, when his medical course was completed, Hayden moved into Hall's "Office" to share quarters with Meek. Here he found temporary work to pay for room and board, meanwhile gaining valuable experience from his association with Hall and Meek, and, as opportunity permitted, writing letters relative to employment. His correspondence was launched with letters to Baird and Leidy written, respectively, on January 8 and 11, 1854. To Baird he wrote:

> At the time of your visit last fall you proposed to me the idea of entering the Army as a Surgeon... Perhaps you may think of some other position in which I could be more useful to Science than in the Army!
>
> I am now preparing myself to make collections, with as much facility as possible [for] the coming season. Prof. Hall wishes me to collect for him but he will not pay me much for it. But some position in an expedition, however humble, would I think be much better for me...
>
> If nothing better presents itself, I may go to Mauvaises Terres for Maj. Vaughan, Indian Agent, if it is thought best. He made me a fine offer. In that case the collection would be under my control and I might do much toward collecting the Reptiles, Fishes, and Insects of that wild country.
>
> If it will not trespass too much on your time I should be glad to hear from you as soon as you receive this, and please give your opinion of what you think would be best for me to do.[3]

To Leidy, Hayden wrote:

> Pardon me for the liberty I take in thus addressing you, but I am anxious to obtain your opinion and counsel in a matter which is of much interest to me. I am very desirous of studying more thoroughly Natural History as well as Medicine & Surgery, and my mind is directed toward Philadelphia as the best place for me to do so. I am anxious to prepare myself for some permanent position at some future day.
>
> Prof. Baird seemed anxious to have me enter the Army as Surgeon. This would be my wish and then I think I could contribute much to Natural History by collections, at least. I would like to ask you how I could get along in Pa. and pursue my studies, if I could make some explorations for the Academy or do anything to defray a part or all of my expenses?... I have graduated from the Medical College here not only poor but somewhat in debt. I cannot endure the idea of giving up my hopes of doing something in Natural History and I am willing to get along most any way to do so. Might I hope quite positively for some position in an expedition next spring? I am now ready to go anywhere on any notice, and I feel very anxious to be in the field at work.
>
> When you were here I spoke to you of the proposition made me by Col. Vaughan, Indian Agent, at Fort Pierre. He made me a fine offer I think. But the question with me is whether I had better accept that or hope for a position in some public expedition. Dr. Engelmann of St. Louis is anxious to have me secure a berth as Surgeon and Naturalist to some government expedition, in preference to a private one. I would like to have your counsel...

> Should it seem best for me to go for Col. Vaughan I would like to make some arrangements with you so as to bring my collection directly to Pa. for I would like to have you examine the Vertebrates & Prof. Baird the Reptiles and Fishes, etc. I am confident I could make a fine collection, much better even than we made the last season, having much longer time. My own expenses would not be much. Prof. Hall has offered me one hundred dollars and takes my collection as security, but I wish my labors for next season to assist me in my studies for the Army or some other berth... I have spent considerable time in preparing myself to collect well and with discrimination in all departments, and now I am giving my attention to Geology altogether, so as to make field notes that may be of some value.[4]

Baird's letter to Hayden has been lost, but the substance of his advice is indicated in a letter that Leidy wrote to Hayden several weeks later:

> Philadelphia, February 28, 1854.
>
> Dr. F. V. Hayden,
>
> My dear Sir,
>
> I desire you to pardon me for neglecting to answer your letter before. So soon as I received yours I sent it to Prof. Baird, and upon its return with one from him I lost both, and could no where find them until yesterday. I have, however, not yet been able to hear of any position under government which you could take.
>
> I send you the portion of the letter of Prof. Baird which has reference to yours. I think with him, if no Government Expedition offers soon, you had better accept the invitation of Major Vaughan...
>
> With respect I remain at your service,
>
> Joseph Leidy.[5]

Hayden's next letter to Baird written only two and a half weeks after his first contains the fullest statement of his tentative arrangements with Vaughan. Very evident in this letter is his disaffection toward James Hall and his determination to terminate his connections with Hall at an early date:

> Albany, Jan. 27th, 1854.
>
> Professor Baird,
>
> Dear Sir,
>
> I received your kind letter in due time and feel under much obligation for the trouble you took for me. I hope I shall yet be able to do something for you, by way of collections, in return for your kindness.
>
> I look upon the Army as the principle object of my efforts though my age will permit me to wait 3 or 4 years yet if necessary. In the mean time, should any position present itself that I might occupy for a year or two perhaps it

would be better for me to accept it, would it not? As an expedition to West, for instance...

I will explain to you as far as possible Col. Vaughan's proposition, his object and plans. He is Indian Agent to the Upper Missouri tribes of Indians and has his station at Fort Pierre. He is a man of considerable vitality residing at St. Joseph's, Mo. and owns much property there. Has no family. Has been Indian Agent among Potawatamies for 8 or 9 years. He has a good deal of energy and decision and I have no doubt will do as he agrees.

I became acquainted with him on the Fur Company's boat last summer on our way up the River. He proposes to pay all of my expenses from St. Joseph's Mo. (his place of residence) and give me half of all I can collect in one year's time. He gives me an ample outfit and all the facilities his position can afford me (which as you know would be very fine) and wishes me to explore not only the "Bad Lands" proper, but also the Bad Lands of the Upper Missouri as far as Falls of Mo. even to Fort Benton if possible. You can easily see that this would afford me one of the finest fields in the world and I have not the slightest doubt but that I could bring as large and valuable [a] collection as was ever brought from the West. It is a glorious country for a Naturalist. Every step he takes, there are multitudes of objects to excite his enthusiasm. I long to explore it thoroughly, every nook & corner. I need not tell you (although the collections from there have been very rich) nothing has yet been done compared with what there is to be done. All the collections from the "Bad Lands" have been made in a circuit of a few miles, while they cover a surface 100 miles in length to 40 or 50 miles in width. There is no reason why further investigation should not develop other localities of still greater interest. Besides, the illustrations in Catlin's work[6] and in the Prince de Wied's magnificent journal[7] seem to indicate the same kind of broken country in the region of Fort Union to Yellow Stone. In fact, Mr. A. Culbertson says they possess still greater interest (though that can hardly be possible). At any rate, it is a grand country for a Naturalist.

That Col. Vaughan was sincere I have not the slightest doubt. He says he has thought of the plan for some years but had not before found a person who suited him to make the offer to. He asked Dr. Evans' advice and Dr. E. told him that it would be a fine idea for both of us and that he could not have a better collector than myself in the country. This aroused his enthusiasm very much and he took me into his room and wanted me to obligate myself to return in the spring, just before we left Fort Pierre, and offered me every inducement. He is willing to expend about 1500 dollars and even more if necessary. But I think that sum will be amply sufficient under his auspices.

You speak of giving the expedition somewhat of a semi-official character, that would aid it much. They are well acquainted with Prof. Baird's name in that region, and with them it is synonymous with the Smithsonian Institution. Anyone going up there under your patronage would be most favorably received and aided in every way. Should some other expedition be fitted out the coming season, for that region, it would of course supersede the necessity of my going, as alone. In this case, I should like to form one of the number...

I am expecting to hear from Col. Vaughan any day. He is probably on his return and will be in Washington in the spring. I wish you would consider this plan and let me know at your earliest opportunity what you think of it, and what additional features you would like to give it. If this plan does not seem best to you I hope some other will occur to you by which I can still further pursue Natural History.

Any thing you may see fit to do for me, will not interfere with any plan of Prof. Hall's for my obligations to him will most certainly cease by the first of March or sooner. He has proposed to give me an 100 dollars for my personal outfit and take my collection for security. But no collections of mine shall ever go East. He held out great inducements to me last spring to do well for him. I did so. That ends the matter. I refused a position in a public school, worth 600 dollars, to go for him, thinking from what he said it would be the stepping stone to fame & fortune. For my hard and faithful services to him last summer, I have received about 75 dollars, and I shall leave him some 50 or 75 dollars indebted to him. Whereas the conditions on which I went were enough to relieve me from all embarrassment. I placed the utmost confidence in him. Perhaps he has done perfectly right. I will not complain. But I cannot stay in this part of the country much longer than March.

You ask me if I know anything about Surveying, Dr. I do not, practically. I am thoroughly acquainted with all the higher branches of Mathematics, and I have Yankee enough to fit myself in most anywhere. I have already obtained considerable knowledge of Geology and Paleontology. I do not try else at present, but that — examining fossils & I have made myself quite familiar with Prof. Hall's collections as far as described. Mr. Meek & myself are now taking casts of fossils. We succeed so well as to receive the commendation of Prof. Agassiz. I am also acquiring the art of sketching & drawing fossils. I can already draw sections & maps, etc. so that I might make myself generally useful in a Survey...

I remain ever yours with much respect,

F. V. Hayden[8]

Hayden's final letter to Baird from Albany is incomplete, but some extracts from the pages extant are worth quoting:

Albany, Feb. 8th, 1854.

Professor Baird,

Dear Sir,

Your last letter encourages me much. I think something may be done at least toward exploring that interesting country. I started a year ago with the most perfect enthusiasm, resolved to give body and mind to the pursuit of Natural History — regardless of all the comforts of life — but for a few months past I have felt much discouraged.

I have not yet heard from Col. Vaughan. He must have returned by this time. Doubtless Mr. A. Culbertson would know something of his whereabouts. He wished me to write him at every opportunity. I now have several letters on the way to him, but they may have failed to reach him. If you can find out where he is, I wish you would inform me where he is and I will write another letter to him immediately. He must be in Washington soon. A few words from you on the importance of such a plan as he proposes would set him right…

Your plan of going across by Minnesota I like very much. This would give me a fine chance to collect Fishes, Crustaceans, etc. As near as I can find out, departments in Zoology which contain very many new species [are not to be found?] west of the Mississippi. I went down with a small hand net only once in Sage Creek, near Bad Lands, and caught 6 or 7 specimens of fish, and Prof. Agassiz finds two new species. I desire much to make a fine collection of fishes, etc. from that region.

The Buffalo, Big Horn, in fact all the mammals peculiar to that region are fast passing away. I would like to make complete collections of certain skeletons of these animals — or at least good heads — and send them into the States, if it would be worth a while. If I remain up there during the Winter such a thing should be possible…

I expect to leave here soon for my home, Rochester, N. Y., probably not to return again. When I leave I will inform you so that you may know where I am, should anything turn up. Prof. Dewey of Rochester is somewhat interested in me and will give me a chance to study while I stay in that place.

Please advise me what arrangements I had better be making — how soon I can start either route by Minnesota or Missouri River — when I had better come to Washington, etc. I feel very anxious to spend as much time at W. and get as many valuable hints as possible.

I am now making my studies useful to me in Nat. Sci. but I could learn many things at W. which I could get nowhere else.

I remain with much respect,

Yours,

F. V. Hayden[9]

Hayden did leave Albany, even before the end of the month, and on February 25, 1854, he wrote Baird from Rochester, giving his new address and inquiring if a collection of fishes from the mouth of the Genesee River and Lake Ontario would be of interest.

With Hayden's move to Rochester, he and Meek became separated, and there began the voluminous and increasingly intimate correspondence between these two friends. Meek's letters, as compared to Hayden's, tended to be dignified and formal in tone for a long time. The following is a portion of Meek's first letter to Hayden in 1854:

Albany, March 4th, 1854.

Mr. Hayden,

My dear Sir,

I received a few days since a letter from Prof. Swallow asking me if I knew of any active young man he could employ as a collector of Botanical specimens in the Missouri Survey. He said he would prefer a young man who was interested in Geology or at any rate one who would like to study that science. He says he will pay to such a person one dollar per day and expenses. I have since written him and recommended you. I let him know your address and I suppose you will soon hear from him. If you have not made other arrangements I would advise you by all means to accept his offer should it be made. It would give you a foothold in western Surveys, and should the state go on with a general natural history Survey your chance for the Botanical department would be better than that of any other person, and at the same time you would have a fine opportunity to pursue your studies in geology etc.

I think I have said enough to make all things smooth with Prof. Swallow in the letter I wrote him…

Very Respectfully etc.

F. B. Meek

P. S. I have just seen a copy of a geological map sent on by Marcou to one of the Bureaus at Washington in which he colors a large district between the western boundary of Arkansas & the Rocky Mt. Range as New Red and Oolitic!!! He considers a large portion of country from which Roemer collected many of his Cretaceous fossils and all that on the headwaters of Red River from which so many Cretaceous fossils have been collected, New Red. Wonder what this remarkable genius will discover next?[10]

James Hall also wrote Hayden, on March 1, 1854, urging consideration of the position with Professor Swallow. He stated:

…I think he will make no objections on the score of your having been engaged for me. On the other hand I think it will be a recommendation to you. I have no idea that he or any one else at the West cares to maintain open quarrel with me, and if they do they shall have the quarreling all on their side.

If I were in your place I would not hesitate & accept such a place as this. It serves to make a beginning and every step afterwards will be easier.

Instead therefore as you feared that your Mauvaises Terres expedition [might be] unfavorable to you [in] the West I think it will introduce you favorably among them. Let the dead past bury the past and begin with them as if nothing had happened that left any unpleasant remembrances.

In this letter, Hall also called Hayden's attention to another opportunity, one evidently resulting from some inquiries he had made on Hayden's behalf:

I have received a letter from Mr. Greeley[11] in which he says, 'News from Nebraska will reach us sooner by the St. Louis papers than by your friend's letters; but if he will write us one letter weekly thence — say from one to two columns in one letter, we will pay him $5 for each letter we print. We should expect to hear not only Topographically, Geologically & Botanically from the new Territory but also as to settlements, emigration and the prospects of its being surrendered to Slavery.'

You can judge whether this promises anything worth considering. If you were there with other objects it might occupy you rainy days.[12]

Writing to Meek on March 4, 1854, Hayden stated:

As my feelings are now, I am willing to do anything to get a start... I have as yet heard nothing from Col. Vaughan, and the prospect is poor of getting an appointment in a government survey. I am very well now but in no business and shall not find any here I think.[13]

In this unsatisfactory state of affairs, Hayden — who quite obviously could not stand inactivity — wrote Baird on March 5th as follows:

Had I better relinquish all hope of going to "Bad Lands" for Maj. Vaughan and a government appointment and accept the offer of Prof. Swallow, or wait for a better opportunity to present itself? I am hardly able to decide. Our past correspondence leads me to look to you as my best and safest adviser, in matters of science. If it will not be asking too much of your kindness, I would like to have you advise me what I had better do in the matter.

I have not yet heard from Maj. Vaughan but cannot but believe something will grow out of his proposition yet. I have labored hard to prepare myself to do good service the coming season and my love for Natural History is so great that I hardly feel any disposition for anything else. But thus far instead of supporting myself I have run behind, and my relatives (who can appreciate nothing which does not bring in an immediate return of cash) strongly oppose my having anything more to do with it... My relatives oppose my going to Mo. on such terms, but if it seems best, I will go and work bravely and trust to the future for my reward.

They are now catching fish quite extensively... a few miles from here. I shall spend some time there the coming week and will secure a good collection for you from that locality. Soon they will draw the seine in the River and at the mouth of Genesee and [I] shall keep close watch. I cannot search the small streams before April.[14]

Two weeks later, indirect news from Vaughan sent Hayden's spirits soaring, and with renewed hopes he again wrote Baird:

Rochester, March 20th, 1854.

Prof. Baird,

Dear Sir,

I have just seen a notice in the *New York Herald* of March 18th which gives me much pleasure and hope. It quotes from the Council Bluffs [Iowa] *Bugle* that Col. A. J. Vaughan has returned from Fort Pierre. He was at Council Bluffs about 1st March and is now probably at St. Joseph's, Mo., his home, unless he is on his way to Washington.

I now feel the strongest hopes that Col. V. will be ready to carry out the plans he proposed to me last season. The paper speaks of his having collected specimens of various petrifactions. The interest felt in these will give him new enthusiasm. He is a man who is fond of being much noticed and looked up to, and a few words from you would set him right were he of the notion. When he kept repeating his offer to me and he being a stranger to me and I distracted him somewhat he took me aside and said I might depend on him most implicitly, that he would do just as he said — that it was not the excitement of the hour. He said he had had the plan steadily in view for years (for you may know that he was Indian Agent among Potawatamies 9 years) and was only waiting to meet with a suitable person to employ. My industry and enthusiasm took his fancy and he wanted very much to bind me in some way to enter into it that coming season. My inexperience made me hesitate, for which I regret.

You can see the prospect! I am so sanguine that I can see nothing but success. It is the darling wish of my heart to go into that wild country. If he can be induced to carry out the plans he proposes, Science will reap a good reward from it, and as you said in a former letter, the past shows larger results for Science from private enterprise than otherwise — that is in proportion to the expense...

I desire very much to come to Washington to get an idea of your way of preparing specimens — for I wish to pay good attention to Zoology, etc.

I have made no permanent arrangement with Prof. Swallow and shall not for some time to come. I do not think it possible to make the arrangement you speak of in your letter. My friends think that I had better wait for something better.

I am greatly obliged to you for your kind offer in your last, but think I shall not need it unless in some scientific enterprise. I have thought of this plan — to go to Council Bluff and then take sketches and gather materials for a History of Nebraska. I have no doubt a book would sell well when the territory is organized. A New York paper has offered to give me $5 for one letter a week from that region, which would help some. My expenses while there would be nothing and I calculate that $100 would enable me to accomplish that object and do something for Science. P. A. Sarpy of Bellevue, Nebraska, offered to hire me any time and give me a chance to explore. They are anxious to have as much said about their place and country as possible and would aid in any way.

I am now spending most of my time with Prof. Dewey at the Rochester University. They have 2 large rooms filled with objects of Natural History in the most beautiful confusion. Hardly a name or a label. I am now assisting Prof. D. to arrange it and name the specimens, which affords me a fine opportunity to pursue my studies.

Wherever I may be, I shall lose no opportunity to collect those objects in Zoology which you speak of and forward them to you. I hope to get a good collection for you from this region.

But my main object now is to find out Maj. V.'s mind, which I shall know in a few days. I have just written him at St. Louis.

I wish much to start as quickly as possible and go up by land — from Council Bluffs or some other point. I would like to send down something by the Boat on its return…

I remain ever yours with sincere regards,

F. V. Hayden[15]

In a letter dated March 21st, Meek wrote Hayden as follows:

I forwarded this evening to you a letter which I suppose is from Dr. Engelmann as it is mailed at St. Louis and directed in his hand. Prof. Hall rec'd a letter from him this morning in which he says he has written you. He says Culbertson has made a treaty with the Blackfeet and that Col. Vaughan has only just arrived at St. Jos. Col. V., Dr. E. thinks, will be the Governor of Nebraska. If this should be the case, and he is willing to go on with the arrangement he made with you, I think after all if I were in your place I would prefer it to the position in the Missouri survey. If Vaughan is willing to take hold of the matter in earnest and should be appointed governor of Nebraska, you could have every facility for pushing your explorations far into that interesting and little known region beyond the Black Hills. You are yet young, and if you were to devote even one or two years of your life to such an enterprize I think it would be the best thing you could do. The treaty with the Blackfeet will make them more pacific, so that under the auspices of Col. V. and through the agency of Culbertson I think an expedition could be fitted out so as to explore the country with perfect safety. I suppose you are aware that the Nebraska Territory as now understood embraces also the vast region between the Black Hills and the Rocky Mountains and extending north to the British possession. It is high up in this region where the Bad Lands of the Missouri, of which Culbertson so often spoke, extend over a vast area…

I think Prof. Swallow wishes to make a permanent engagement, as he said he would like to employ a young man who wishes to study geology. Of course he did not expect any person to accomplish much in this way in a few months.

I hope you may soon be able to make up your mind for that which may be most to your advantage. Let me hear from you soon…

I wrote Swallow & Shumard both strong letters for you…

I found one of your bows today in moving one of the specimen cases. I have seen nothing however of the bow case or quiver. Shall I send the bow by express?

F. B. M.[16]

When Hayden replied to Meek, he was still sanguine as to the prospects with Vaughan:

Rochester, March 27th, 1854.

Friend Meek,

I received your letter with much pleasure and hasten to reply. As yet I have heard nothing from Prof. Swallow and do not much expect I would suit him...

However, the prospect is now that Maj. Vaughan will perform his promise, and this will lay all other things aside... He is now probably at St. Louis — was at Council Bluffs 1st March — came from Fort Pierre by land — 21 days. No snow in the mountains of any account. There will be no rise in the River this Spring as usual...

I shall make the plan go, if there is the barest possibility.

Prof. Hall is kind in offering to furnish me an outfit but I think he is rather too poor. He is too close a man for me were there no other reasons...

I have spent my time most agreeably at Prof. Dewey's, studying the Natural History. He thinks a good deal of me and gives me a good word whenever he can. He would like to see you very much and if you ever find [yourself out] this way you must call on him. There are a vast number of fossils in the University collection and these we are studying out and labeling. He gives me duplicates and I have made a fine collection in this vicinity some of which are as good as any Hall has...

I am glad you have found that bow. Please send it by express as soon as possible. Is it my nice ornamented bow? I hope you will find the case [and] quiver. If not do keep a lookout so that at some time you can give me a history of it...

Direct the bow to me Rochester, New York, and [I] will get it. Please write me as often as convenient, and believe me most truly and sincerely your friend.

F. V. Hayden

I am much obliged to you for writing to Swallow & Shumard.[17]

Meek's next letter is noteworthy because in it he first proposes collaboration in the study of any new Cretaceous fossils which Hayden might discover:

Albany, Mar. 29th, 1854.

My dear Sir:

Yours of the 27th Inst. came to hand yesterday. I was much pleased to learn you have a prospect of getting an opportunity to visit the Upper Missouri country again...

If you go to the U. M. country again do not fail to collect all the new Cretaceous fossils you can, and if you can find no other person that would suit you better, I would like to join you in investigating them. I would be very much obliged to you for any fossils you may be able to procure in the region of Rochester. I would like to become acquainted with Prof. Dewey and will not fail to call on him whenever I visit Rochester.

I sent your bow by express yesterday evening. You will doubtless have received it ere this reaches you.

I think you will certainly have a letter from Prof. Swallow soon if you have not by this time. I think from what he said he had you in view when he wrote but feared you were engaged here and did not wish to make you an offer until he could ascertain whether or not you were free to accept it...

Yours sincerely etc.

F. B. Meek[18]

In early April Hayden received a letter from George Engelmann, and posted the following letter of reply:

Rochester, April 5th, 1854.

Dear Dr. E.

I received your welcome letter with much joy... I am sincerely obliged for the trouble you have taken for me.

As yet I have heard nothing directly from Col. Vaughan. I think if he is still disposed to carry out his plan I shall accept his offer, and have written several times signifying my acceptance...

I will enclose a note to Col. V. which I wish you would be so kind as to get to him some way. If he is in St. Joseph's yet, please send it by mail to him as soon as possible.

I am very glad if there is anything interesting among the Cacti. If I go again I shall be more particular in my search for them. I hope you will obtain a set of the plants. They are all with Prof. Hall.

I ever remain with much respect,

Yours,

F. V. Hayden

If I fail to go for Col. V. I hope I can go with Culbertson to Blackfoot country. I will go most any way and for most any price or for my expenses simply...[19]

After the months of waiting, Hayden finally did receive word from Vaughan, and it was negative. This unhappy news Hayden relayed to Baird on April 11th:

Rochester, April 11th, 1854.

Professor Baird,

My Dear Sir,

I have just received a letter from Col. V. which I will enclose to you. I regret to say that it is unfavorable... What can be done now? I feel the deepest interest in that country and it seems to me that I cannot give up the idea of making further explorations there. I am now quite impatient to be on the ground at work so as to secure as large a portion of the season as possible. I do most seriously hope something else will occur to you to render our plans practicable yet. Anything that may seem best to you will be perfectly satisfactory to me. I am ready to go for my expenses simply though I am not able, yet rather than have the plan fail I will do most any way... Dr. Treudeau was physician for the company last year, but died soon after his return to St. Louis. Perhaps I could get that berth by a lively application. Mr. J. B. Sarpy who went up in charge of the boat last summer was very kindly disposed toward me, and repeatedly urged me to become a trader. I hope some other way will open — still any way to accomplish the main object that is honorable.

I received a letter a few days since from Dr. Engelmann expressing a strong desire that I should go into the Blackfeet country with Culbertson. He said he should write to you soon on the subject. I suppose he has before this. You spoke of it in your last. Can anything be done in that direction effectually? Could I not be attached to him in his official capacity in some way and thereby lighten the expense — for I suppose the expense is the main barrier in the way of accomplishing our designs. I do hope the Fur Company's boat will not go up without carrying some one who will feel an interest in developing the Natural History of that country. With Mr. Culbertson I could make a small sum produce fine results, for he is a man of liberal feelings and has an interest in Science...

I have been very busy since I have been here, with Prof. Dewey labeling the collection of the University. We are soon finished. I have also made fine collections in this vicinity...

I remain ever yours with respect,

F. V. Hayden[20]

Meanwhile, G. C. Swallow offered Hayden a position with the Missouri Survey:

St. Louis Mo. Apr. 6th '54.

Mr. H. V. Hayden

My Dear Sir:

I write you by the suggestion of your friend Mr. Meek, to offer you a place in our Survey. Your duties will be to collect plants & fossils principally.

You will be with me the most of your time. We shall want you till Dec. next, & longer should the Survey continue. We can give but $1. per day and pay your expenses while in the field & all travelling expenses after you reach St. Louis. Come as soon as you can if you conclude to accept our offer, and let me know at Columbia Mo. whether you will or will not & when should you conclude to come.

Yours truly

G. C. Swallow[21]

By mid-April, 1854, Hayden's prospects still remained unimproved, and he reconciled himself to accepting the one position open to him, that with the Missouri Survey. Before starting for Missouri, by way of Cleveland, he wrote Baird as follows:

Rochester, April 14th/54.

Prof. Baird,

My Dear Sir,

I have just received your kind letter with much pleasure, though I am somewhat disappointed at not being able to visit the West this summer.

I do not think much can be done at Cleveland. I think Dr. Newberry would not feel able to do anything. Prof. Kirtland might do something. He is always ready to lend a helping hand in Natural History.

I received a letter a few days ago from Prof. Swallow requesting me to come on to Missouri as soon as possible. I have concluded that the best thing I can do now is to go to Mo. He can offer me but one dollar per day and expenses while in the field after arriving in St. Louis. You can readily see that this is but a small pittance. It will cost me at least $25. to go to St. Louis and what my other expenses will be I cannot tell. However, as nothing better seems to offer I will go and work earnestly — trusting that the future will open some way for me to do better... I shall have it understood that I can leave if I should have an appointment under government.

I see that the Gadsden Treaty[22] has been rejected. I am much pleased with the idea of Captain Marcy's Expedition. Could I obtain a position in such an expedition it would give me the greatest pleasure. I hope it will become a certainty.

Although prospects do not now look as bright as could be wished I have put my hand to the Mo. plough and cannot turn back...

I expect now to start for M.[issouri] Monday. Should you hear anything important please direct to C.[leveland] care of Dr. Newberry. I must stay in C. two or three days.

I shall spare no pains to collect Fishes, Reptiles, etc. for you. I do not think Prof. S. considers them foreign to the Survey. Prof. Dewey has been examining the Carices that are in the collection of plants. He finds three new species, which he has described. I think if the others could be examined there would be other new species.

I do hope that Captain Marcy's plan or a similar one will find favor. That region has been one of the greatest interest to me and I should love very much to search out its Natural History. I think your advice about my going to Washington much the best. I am rejoiced that the way is opened for me to work at least, and I shall be better prepared to examine more thoroughly wilder regions. I hope I may hear from you again in Mo.

I shall ever remain with the kindest regards,

Yours,

F. V. Hayden[23]

Hayden's departure from Rochester was delayed, and fortunately so; for in the meantime Vaughan arrived at Washington, was interviewed by the persuasive Baird, and, as a result, agreed to carry out the plans originally discussed with Hayden. Baird lost no time in acquainting Hayden with this favorable turn of affairs. In letters dated April 25th and 26th, Hayden expressed his elation over the good news. In the second of these letters he also touched on his purposes and plans, and inquired about facilities which he would need:

Rochester, April 26th, 1854.

My Dear Sir,

...I need not say how much I am rejoiced at having an opportunity to make further explorations in that wild country. Once in the country, I think I could stay without great expenses. The opportunity is too rare to be lost under any consideration. Science demands that it should be improved, and although I have accepted Prof. S.'s offer I am confident that if he feels any interest in Science he will not hesitate one moment to release me from any obligations to him, especially as he can find enough who would fill the place he wishes me to have, as well as myself.

I shall be able to make fine collections all the way to Fort Union, and then if the goods are transferred to boats, these boats must be "cordelled" or pulled by hand, for 250 miles up the Yellow Stone. All of which country I can explore thoroughly for the boat cannot go more than 15 or 20 miles per day. I can see nothing to hinder me from making one of the finest collections ever brought from the West.

Should I go up to spend a year I should need an extensive outfit. I suppose you will be able to furnish many things and I doubt not that Col. V. will make allowance for them. I shall feel like husbanding resources as much as possible.

I shall need as correct a map of that region as I can get to paint down the geology as I determine it. Would not some cheap portable barometer be good? Also a thermometer & instruments of that sort. Can you send me a volume printed in England for the Naval Service — edited by Sir John Herschel — giving directions in making collections? I liked that very much. Or some other book. I shall need $50 or $100 to make an outfit, aside from provisions. Will Col. V. permit that? Dr. Engelmann will be ready to assist as usual.

Would it not be best to make an arrangement with Col. V. to have all the collections sent to Washington and examined before any division takes place?

I have already written to Col. V. at New York. I do not know that it will be necessary for me to go there. I do not doubt but that I can persuade him to furnish abundant means when once started.

I thought I would start for Cleveland next Monday or Tuesday. I will write you again from that place. I wish to go through C. as I have many things there which I shall lose unless I go and take care of them.

I hope as good preparation may be made for collecting specimens in Zoology as possible, for I wish to pay particular attention to it, especially Fishes & Reptiles. I hope you will feel the liberty to identify yourself with the whole enterprise for as far as I am concerned I wish to submit the direction and disposition of my part (at least) to you.

I shall ever feel in the greatest degree indebted to you for your kindness and I trust that if I live to return I shall bring back sufficient additions to Natural History to repay you for your trouble. I wish very much to visit you at Washington but there will be no time now to make it worth a while. I will meet Col. V. at St. Louis in due time and all things can then be made right.

With great regard, yours,

F. V. Hayden[24]

Baird's intercession on Hayden's behalf set matters moving swiftly and decisively. A few days after the above letter was written, Hayden made a hurried trip to New York City for a conference with Vaughan, presumably at the latter's request. The outcome of their talks was encouraging, judging from a letter which Hayden wrote to Baird:

New York, May 5th, 1854.

Prof. Baird

My Dear Sir,

I have spent a few days here with Col. Vaughan and everything so far is satisfactory. He agrees to let me explore about 18 months and perhaps longer if successful. He seems willing to do every thing fairly. Still our bargain is merely verbal and I would like to have you see him when he comes to W. and talk the matter over with him. He has a high opinion of you and your influence with him will be great. If he states our bargain to you I shall have proof and he will fear to do anything less than he agrees. I do not know that this is necessary. It occurred to me however.

Col. V. starts for W. tonight. I can see no reason why the enterprise will not be successful and for your efforts in bringing the matter about I am very grateful. I will do the best I can. I trust you will continue to feel a deep interest in it, the same as though it were at your own expense, and I shall feel a great pleasure in giving you of its fruits what you may desire. I shall need some funds at St. Louis for my own personal outfit which I cannot obtain otherwise than of you. I will secure you in any way you may say, trusting that all will be right in the end.

Col. V. pays all expenses from St. Louis back to that point. All other expenses I must bear myself, which is probably fair enough.

I wish you would send to St. Louis any reports on Indian affairs which may be useful to me. During the winter and on the boat I shall have plenty of time to use them. It will be lonesome to me without a plenty of books and any that you can send me on Natural History will be acceptable... Any instruments, or apparatus, that you can send by Col. V. or by any other means will be very much needed. For preserving reptiles and fishes I wish to be well prepared. Col. V. may think some things unnecessary as he knows nothing about these things. Should this be the case I wish to get them at my own expense rather [than] neglect anything that will in any way cripple my efforts.

I deeply regret that I cannot spend a week with you for I could get many hints that would be very useful to me...

I hope you will freely advise me in all things that I should do and if possible send me a list of articles needed.

I trust I shall be able to add many things to Science. I remain ever most truly & sincerely yours,

F. V. Hayden[25]

After Vaughan had revisited Washington, and again conferred with Baird, the latter methodically filed away in his records the following memorandum:

Dr. F. V. Hayden of Rochester, New York, goes in May 1854 to the Yellowstone River in company with Col. Vaughan, Indian Agent, who takes up goods for the Crows and other Indian Tribes. Col. Vaughan authorized me to tell Hayden he would give him 300 or 400 dollars or more if necessary, see that his expenses were reduced to a minimum, and let him keep half of the various collections in Natural History which he may make. Subsequently Dr. Hayden informs me by letter that Col. V. agrees to pay all expenses from St. Louis and back. As Dr. H. requires an outfit and money to meet necessary expenses I have applied to the following persons for subscriptions to be repaid in specimens

Prof. J. P. Kirtland, Cleveland.	20.00	Paid. C ? elae & Birds
Prof. Chester Dewey, Rochester.	10.00	Promises to send to St. Louis.
I. H. Clark, Phila.	10.00	Paid.
J. L. LeConte	10.00	Paid. Small Coleoptera
Rev. Chas. Fox, Detroit	10.00	Paid.
Dr. T. M. Brewer, Balt. (?)	10.00	Paid.
	60.00	collected

May 11, 1854. S. F. Baird[26]

Another memorandum pertaining to equipment that Baird furnished to Hayden was made the same day:

May 11, 1854. Sent to Dr. Hayden,

1 Leather Trunk	Belonging to Lt. Whipple. Lent by him.
1 Copper Trunk	"
1 Inflating bag	"
1 Pair Scissors	"
1 Iron Wrench	"
1 small seive	Sm. Institution
2 cans Arsenic	"
1 Piece Line	"
Blank labels	"
Parchment	"
Various miscellaneous books.	

About three weeks later, Baird added to the foregoing memorandum a notation stating, "June 2, 1854. Sent draft for $60.00 to Dr. care Engelmann ..."[27]

After his rendezvous with Vaughan in New York City, Hayden returned to Rochester, and just before starting west he wrote Baird a brief letter:

Rochester, May 13th, 1854.

Prof. Baird,

My Dear Sir,

I start today for St. Louis. Shall spend the day tomorrow in Cleveland, and go by the way of Chicago. The boat, I think, will start as soon as the 20th...

I received a letter a day or two ago from Ann Arbor, Mich. from D. Douglass requesting to know my terms as an assistant... Col. Vaughan promises very fair and if he fulfills I have no hesitation in prefering [*sic*] to go with him. I think there will be no difficulty in getting released from Prof. Swallow. My engagement with Col. V. was long prior to any with Prof. S. I shall be sorry to disappoint anyone or cause any hard feelings... and I wish to take no wrong step in the matter.

I shall need $50 or $75 for my personal outfit which I desire you to furnish me at St. Louis. I would like the sum of $75 to draw on though I shall not need as much, I think, but in case I should it would be pleasant to have it.

Our understanding was that I should spend a year at Fort Benton and Fort Union, and then spend the Spring until the boat returns, one year from next July, at the "Bad Lands" and around Fort Pierre. This will give a fine opportunity to make large collections and should be improved it seems to me.

I hope you have been able to send some apparatus to St. Louis. I suppose Col. V. will not feel willing to spend more than $100 for an outfit.

Prof. Hall has written to me desiring to let me have $100 and take my collection as security, but past experience tells me that I could not do so safely.

I remain ever yours,

F. V. Hayden[28]

On the same day, Hayden also wrote a short letter to Leidy:

Rochester, May 13th, 1854.

Prof. Leidy,

My Dear Sir,

I received a very kind letter from you some time ago. I also received a copy of your "Fauna of Nebraska" for which I desire you to accept my warmest thanks.

I was in Philadelphia a few days ago but was sorry to find you were not in the city. I wandered into the "Academy" perhaps improperly but I did no harm for I did not touch anything not even a book.

I suppose Prof. Baird has informed you of my intentions of going to the far West this summer. All things appear well at present. I trust some good will come out of it…

If you can give me any suggestions respecting collecting Mammalia I should be glad to have you do so, for fossil and living Mammalia will be my main object. The boat starts from St. Louis May 20th.

I remain ever

Yours,

F. V. Hayden[29]

To this letter, Leidy replied:

Philadelphia,
May 15, 1854.

Dr. Hayden,

Dear Sir,

I am sorry I did not see you when you passed through our city. At the time I was in St. Louis as Dr. Engelman [*sic*] may inform you.

When you collect mammalian fossils remember, with limited means of conveyance, etc. it is best to neglect all fragments except those of skulls, and perfect articular extremities of bones. Of reptilian remains vertebrae are very characteristic. If you go to Nebraska, avoid heavy fragments of bones without they are decidedly and characteristically perfect. I would also avoid the heavy turtles, and fragments of the same.

All living mammals from the regions you propose to visit are of interest, but more especially the smaller ones: bats, mice, squirrels, etc.

When in St. Louis Dr. Shumard showed to me a number of slabs of bluish limestone with minute fossil fresh water shells: physa, limnea, planorbis, etc. They are converted into Calcedony. They were found in Nebraska in the upper layers of the ossiferous formation and may be readily overlooked. It would be well worth your while to visit Shumard; you will find him an excellent fellow.

If I can be of any assistance to you, let me know it.

With respect I remain at your service,

Joseph Leidy[30]

On May 17th, Hayden arrived in St. Louis again, where he was cordially received by acquaintances of the year previous. With the aid of Vaughan, he outfitted himself for the journey. He was not pressed for time, as now the boat was not scheduled to leave until June. Before its departure, Hayden wrote Baird two letters:

St. Louis, May 19th, 1854.

Prof. Baird,

My Dear Sir,

I arrived at St. Louis on the evening of the 17th — as I find in good time certainly, for I learn that the boat will not start until some time in June.

As soon as I got here, I visited Dr. Engelmann and was very glad to receive your letter...

The prospect of success seems as far as I could wish. Col. V. is very much elated and everything seems to conspire to keep him so. He has requested of me a list of outfit which I have just made out...

When I left home, my friends felt very sadly about my going so far. They had felt so for a week or more, and I did not come away with as high hopes and confidence of success as I had expected to feel. My friends have never seemed to care so much about me before. But when I reached St. Louis and was greeted so cordially by Mr. Sarpy, Culbertson and others, I began to feel that it would be a pleasure to me to spend two years even, exploring that wild country. They all seem to feel interested in helping me along, and seem to think my industry will make my future a bright one.

Your great influence with Culbertson is of the greatest benefit to me, for there is nothing he would not do for you and now the same kindness will be extended to me, as he is acquainted with your interest in this affair.

I stopped a day or so with Dr. Newberry at Cleveland. He treated me with great hospitality and felt a deep interest in my success.

I am very glad that $40 has been raised for my personal outfit. I am inclined to think that I shall not use even as much as that, though I shall need it when I return. My personal outfit will not cost much. Prof. Dewey gave me $10 which I spent in coming here. My expenses will be something on account of the boat delay. I shall get a cheap boarding place and economize as much as possible. I think Col. V. should pay a part at least of my expenses after the 20th.

Respecting the directions given in your letter, I shall have them carefully in mind and act upon them. Should anything more occur to you please write.

There was much stormy weather this spring and the streams were so much swollen that I could not do enough in collecting for you about Rochester to pay the trouble of sending to you. I have, however, requested my two brothers who take a good deal of interest in the thing to keep a jar of alcohol standing somewhere in a safe place and drop in whatever they may catch. They will accumulate considerable in course of the summer. I requested them to leave the specimens with Prof. Dewey for you.

I received a letter from Mr. Meek here in which he informs me that he has received an offer of a $1000 a year and expenses with a prospect of $1500 the coming year. I do hope he will be released for he is a noble fellow and deserves a better fate. His whole soul is wrapped up in Geology & Paleontology, and in a Survey he would be one of the most useful men that could be found.

I was very much obliged for your kind offer of a month or two ago, but then I could see no prospect of being able to pay you again and as I could possibly get along without it I felt a delicacy in accepting it. It gave me the most perfect confidence, however, that you were my true friend and felt a true interest in my welfare.

As yet I have heard nothing from Prof. Swallow though I have written him stating my intentions to go up the river. There will be no trouble, I think. At any rate I have made up my mind what to do.

I will write to you often and hope I may hear from you again soon.

I remain with great respect,

Yours,

F. V. Hayden[31]

Hayden's second letter to Baird was written six days later:

St. Louis, May 25th, 1854.

Prof. Baird,

My Dear Sir,

I have received your last letter, also the Dispatch. I could not well use your order as Dr. Engelmann had no spare money by him. The Fur Company would have advanced me the money if I had urged the matter, but as I do not go up in their boat I did not feel like asking many favors of them.

They are extremely kind, however, and offer every facility. Mr. Sarpy has offered to carry up any freight necessary for the expedition free of expense. So that the prospect is fair for doing a good deal with small expense.

I am well pleased with the way everything works. Col. V. will do more even than I anticipated. He has already made large purchases of provisions and today he told me to get any thing necessary and bring him the bill and he would be ready to pay it. I bought today 6 reams of wrapping paper, 2 small kegs of 5 gallons each filled with alcohol, ordered a plant press made. Should anything occur to you as necessary, write me. You will have ample time for I do not think the boat will start before a week from Saturday.

Do not telegraph again. A letter will reach me sooner by mail and will be much cheaper. Your last telegraph was five days reaching me. Mr. Meek and myself were annoyed just so last Spring, and we found that the mail was more expeditious than telegraph.

Please send Dr. Engelmann a check to the amount of 60 dollars. I do not think I shall need it all, but some money will be very convenient when I return from the mountains...

The collection of books, etc. you sent me gave me great pleasure. They are just what I wished. I find no Indian vocabularies, but large numbers of meteorological blanks, etc.

Without much doubt now I shall spend the winter at Fort Union. If I return in the Spring I shall not visit the "Bad Lands" at all, but I think I have it so arranged that I shall be able to spend six months there next year. I shall not feel satisfied with my trip unless I can bring that about.

I wish you could ascertain to whom Prof. Hall sends a set of the plants I collected last summer. If Prof's Torrey or Gray[32] or Dr. Engelmann get them I hope a catalogue of the species will be made out. I would like to make out a complete catalogue of all the plants that have been found in the vicinity of the Mo. I hope you will write to me any suggestions or advise me in anything that may occur to you further.

I remain with the kindest regards,

Yours,

F. V. Hayden

Hayden added an addendum to his letter a day later:

Since writing to you yesterday I have had an interview with a man from Fort Pierre, named McEllery. He has been in the employ of Fur Comp. at Fort Pierre 9 years and is well acquainted with every part. He has saved about $1000 and is very desirous of embarking in this scheme with Col. Vaughan and myself. I have laid before him a plain statement of the facts as they look to me and shall talk further with him today. He has an idea that a handsome sum of money can be made of a collection from that region.

Although he would be of the greatest aid, in fact the best man that could be got, his influence being great among the Indians, still by what

> I have said I have rather discouraged than encouraged such a thing. Col. Vaughan is very anxious to have him go with us. McEllery is a very good fellow and unable to lose much in any speculations though it would add at least a third to the result of the expedition to have him go. We should of course make an immense collection and if it were saleable should make something handsome out of it. How that is I do not know. As yet he seems anxious to go into the thing and says if the whole thing is a failure he will cast no reflections on me. He will decide soon...
>
> The trunks you sent reached here yesterday safely. Please write soon. Boat goes between 1st & 15th June.
>
> Yours etc.
>
> F. V. Hayden[33]

For some reason Mr. McEllery did not become a partner to the venture. A few days before their departure, Vaughan and Hayden had drawn up a formal contract covering the terms of their agreement. This document they both signed before a witness:

> Alfred J. Vaughan and F. V. Hayden have the First day of June One thousand Eight hundred & fifty four entered into an agreement to develop the resources of the Upper Missouri for Twelve Months from the above date, for which the Said Alfred J. Vaughan obligates and binds himself to the expenditure of Three hundred Dollars for the above object, and all specimens that may be collected by said Hayden the said Vaughan and Hayden are to share and share alike in. The proceeds of the said Hayden's labour is to be conveyed to St. Louis, said Vaughan incurring all expenses whether it exceed the Three hundred Dollars or not as originally understood, fifty or one hundred Dollars being no object to carry the project fully out, and that amount exceeding the Three hundred Dollars it is understood will be defrayed by Vaughan alone.
>
> The specimens collected after arriving in St. Louis will be arranged and classified by said Hayden and an assistant that will be furnished by said Vaughan, should said Vaughan deem it necessary, and then equally divided between said Vaughan and Hayden, but should it be mutually agreed upon by the parties to keep them together and dispose of them in bulk, will be optionary with either one of the parties.
>
> It is further understood and agreed upon that all plants peltries and alcoholic specimens we have mutually agreed to present to Professor Baird, Secy of the Smithsonian Institution.
>
> Alfred J. Vaughan
>
> Signed in presence of
>
> F. V. Hayden
>
> W. D. Dodge A. G.[34]

While he was in St. Louis, Hayden received a letter from Meek, one which he undoubtedly carried with him and consulted on his journey. In it, Meek briefed Hayden on the known geology of the region into which he was going and made a number of suggestions which might prove helpful in the further deciphering of the geologic relationships, pointing out, for example, problems awaiting solution, and key places where significant fossils might be collected or critical observations made:

Albany, May 4th, 1854.

My dear Sir:

On my return from Washington City yesterday, I found waiting me here yours of the 29th inst. I fear it is now too late for my answer to reach you before your departure from St. Louis; but as it may be sent to you by some of the Fur Company's expresses or otherwise I now sit down to give you briefly (for I am too much hurried with my drawings in consequence of my late absence, to do more) a few facts in regard to the Geological status of the Upper Missouri country that may aid you in your investigations. I have not time to attempt the construction of a continuous section along the line of our observations, from the disconnected facts scattered through my note book, but I will give you a vertical section showing what I conceive to be the order of superposition, relative thickness, etc. of the various beds composing Tertiary and Cretaceous formations of that region. Of course you must not expect to find all the beds represented at any one locality; nor their thickness exactly what I put them down, for you know I was provided with no instruments for taking measurements. It is also probable you may see higher up the river, interpolated beds not represented where we made our observations, and those seen by us may at other places assume quite different lithological characters and vary greatly in thickness.

In the first place I would remark in regard to the Tertiary beds of the Bad Lands that where examined by us their lines of separation are too faintly marked to be fully made out in so short a time as I had to devote in that way. I will therefore remark in general terms that the whole formation where we examined it, appears to be made up of a series of whitish and ash colored clays or marls (sometimes having a slight tinge of flesh color) with seams and beds of white limestone in thin layers and soft sandstone or rather loosely aggregated sand and pebbles. There are also, as you will remember, scattered through most of the beds numerous concretionary masses of harder clay which usually, in the fossiliferous beds, envelop organic remains. Whole formations [are] traversed by chalcedony.[35]

The following section which you will perceive, as far as it goes, agrees very nearly with Dr. Evans' and as the Dr. is a very close observer and had much better opportunities for observation than we had I am disposed to consider his section worthy of confidence. I did not see some of the beds mentioned by him, but they doubtless exist at other places.

[Geologic Section]

Beneath the above section comes the Cretaceous, upon which the Tertiary appears to rest unconformably — both formations having a very gentle dip at Bear Creek to S. E.? Although as I before remarked, no one section shows all the following beds, the order of superposition of the various members of the Cretaceous formation appear to me to be as represented on the following page.

[Geologic Section]

The last mentioned bed is the lowest seen by us that could be certainly identified as Cretaceous. It is very desirable to ascertain the nature of the underlying beds. This doubtful point can be best cleared up by carefully examining the base of the bluffs at low water, between the mouth of Vermillion River and Big Sioux and lower down the river. It is my opinion that the sandstone seen near the mouth of Big Sioux and below there is the next underlying rock and that it is the base of the Cretaceous rocks of the Upper Missouri — <u>though I am by no means satisfied of this</u> as we found no fossils but a Pectunculus which cannot be relied upon in a question of this kind, as species of this genus are found according to some authors ranging from the Carboniferous epoch, through all the subsequent formations and still exist in our present seas. Consequently this sandstone, for all we know to the contrary, might be Tertiary. The fact however that older rocks are constantly seen along the Missouri as we come down the river, shows that the dip of the strata must be nearly in a contrary direction to its general course, consequently it is reasonable to suppose that the sandstone seen at the mouth of the Big Sioux and below are older than the lowest of the Cretaceous beds that show themselves not many miles above. If so, they are probably Cretaceous or Carboniferous, and the fact that the same specimens containing the Pectunculus also contains what appears to be moulds of Dentalia, makes me incline to the opinion that it is Cretaceous. Nicollet says he saw beds highly charged with Inoceramus barabina resting upon Carboniferous limestone within a mile of the mouth of the Big Sioux; if he was not mistaken it is near this point that we must look for the junction of these two formations.

Another point to which I would direct your attention should you descend the Missouri in a small boat is Bijou Hill. It would be very desirable if possible to get a section some where in this vicinity of the beds between the top of the gray calcareous marl containing fish scales and Ostrea congesta, and the Tertiary beds crowning the summit of the elevations in this vicinity. The great thickness of strata (800 or 700 feet) between these two horizons, causes me to suspect there are here some interpolated Cretaceous beds not recognized by us, or that some of the beds swell out to great thickness.

The Tertiary beds on top of Bijou Hill I am now inclined to regard as more recent than those at the Bad Lands, since Dr. Leidy has decided the bones and teeth we found there belong to true Ruminants.

I hope you will look carefully for shells, corals and all other kinds of vertebrate remains in both the Cretaceous and Tertiary beds wherever you may go. A very interesting question yet remains to be settled in regard to the Tertiary formation composition [of] the Bad Lands. That is whether they are of Marine, Estuary or Lacustrine origin. I had the pleasure of discussing this question with Mr. Conrad with whom I spent very agreeably, one day on my return from Washington. I had taken with me the specimens from the Bad Lands which I had supposed to be Natica and Hipponyx and which were so regarded by Agassiz. After carefully removing some of the matrix that adhered to the supposed Natica we found a small portion of the shell still remaining, the striae upon which have nearly exactly the character of Helix. This together with the general form of the shell leaves but little doubt that it is a true Helix. Consequently the only evidence we now have that these deposits were made in a sea or even an Estuary is the Hipponyx, and as we are not positively certain that this specimen was found in the Tertiary (the first I saw of it was in a lot of bones brought into camp from the Bad Lands, with which it may have accidentally been mixed from some of our Cretaceous fossils previously collected) the whole matter is still in doubt. From the thin seams of white limestone near the upper part of the section of Bad Lands we have undoubted Linnea planiorbis and Cyclostoma, which are of course of fresh water origin. Upon a review of the whole question I am inclined to consider these deposits Lacustrine. From these facts you may see the importance [of] looking carefully for minute things in both the Cretaceous and Tertiary. Every one of our small species are new whilst nearly all the large ones have been heretofore described.

If you go with Col. Vaughan to Ft. Union study carefully the lignite deposits above Fort Pierre. Collect all the information you can in regard to their thickness and horizontal extent, as well as in regard to the changes produced by their combustion on the adjacent strata and upon the general physical features of the country, keeping also an eye to their economical value for fuel.

I saw Mr. Culbertson in Washington. From what he tells me about the country beyond the Black Hills I do not think you will find organic remains so abundant as in the region of White River. Whatever you may find up there however will more than likely be new. If you should not be very successful up there I would explore the country higher up White River and between it and L'eau qui Court.

I had a very pleasant time at Washington — remained there four days. Came on and spent one day at the Philadelphia Academy of Sciences and one day with Mr. Conrad at Trenton, New Jersey.

I have recently had an offer of an appointment with a salary of one thousand dollars a year and expenses, and a reasonable prospect of its being raised to fifteen hundred dollars. I am now trying to persuade Prof. Hall to release me honorably from my contract with him and have some prospect of doing so. I will not know however certainly about it until he returns from Washington. He will be back tomorrow.

If this reaches you before you leave St. Louis do not fail to remember me to my friends there. Give my respects to Col. Vaughan and do not fail to write me from the far West whenever you have an opportunity to send a letter.

I found your bow case a few days since snugly wrapped in a newspaper under one of the tables up stairs amongst bundles of waste paper of much the same size and appearance. I will send it directed to your address care of Prof. Dewey, Rochester, and I presume your friends will get it.

Yours in haste

F. B. Meek[36]

In later years, Hayden would receive many such letters in which Meek gave wise suggestions and asked searching questions, admonished, criticized, or scolded and shared with Hayden the news and gossip from their common circle of acquaintances in "the States." Meek's letters, some of which ran to ten or fifteen pages, and Hayden's replies, which could be equally copious, reveal the reciprocal relationship between these two men. They became essentially a team in which each complemented the other. Between them there developed a tacit recognition of the fact that working together they could accomplish what otherwise would be impossible, and, accordingly, a division of work came about. The effectiveness of their teamwork is attested by some two score scientific papers published jointly under the names of F. B. Meek and F. V. Hayden, and also by many other papers which, though signed by only Meek or Hayden, give evidence of the influence of each partner on the other.

Meek remained single throughout his life. Dedicated to his profession, he lived a severely simple, even austere, life, spent for the most part working long and hard at his desk. In his collaboration with Hayden, and later with a few other colleagues, he generally assumed the major responsibility for describing the figuring specimens, interpreting their significance stratigraphically and otherwise, and seeing papers through to publication. For this phase of the work he was superlatively well qualified. He became, in George Merrill's estimation, "the most widely known of American paleontologists" with the exception of James Hall. To quote Merrill further, "Had Meek possessed the tremendous physique of Hall or the nervous energy of Hayden, he might have stood alone head and shoulders above most, if not all, of his contemporaries. As it was, he did his best work only in the service of, or in collaboration with, others."[37]

It is doubtful that Meek could have continued, as a long-range venture, the explorations which he and Hayden had begun in 1853, even had he been free to do so. This field work, however, was cut to order for Hayden, who embraced the opportunity, made a life's work of it, and gloried in the experience. To him its difficulties and dangers were a part of the challenge, adding zest to the doing and satisfaction to accomplishment. Thus Meek's younger colleague became, as it were, his alter ego, who undertook the on-the-spot studies which he himself would have liked to pursue but could not.

Year in and year out, and with increased competence as he gained in experience and profited from collaboration with Meek and others, Hayden demonstrated a truly remarkable capacity for getting the essential field work done despite the trying conditions often imposed by working alone and with the meager or makeshift facilities generally available, despite the hardships of travel in a region of immense distances, severe climate, and rugged terrain, and despite the frequent necessity of living off the land and evading Indians. As an explorer-scientist — and, later, as an organizer and leader of government surveys — Hayden excelled, for he was primarily a man of action and a lover of the outdoors. His restless spirit and energetic body required physical as well as mental activity, and he was therefore much less adapted to routine work in the laboratory. In the Meek-Hayden partnership, thus, each man did the work for which he was best suited; neither could have succeeded equally well if the respective roles had been exchanged.

In the correspondence of the period preceding Hayden's departure from St. Louis, the name of James Hall appears but sparingly. There is record of a letter that Hayden wrote Hall on May 14, 1854, but it has been lost. Hall probably replied but if so his letter too is missing. That Hall was offended by Hayden's defection and alliance with others cannot be doubted. Professor Kirtland, in a letter to Baird dated June 27, 1854, included this note, marked Confidential and boxed in with heavy lines: "Do you know that Mr. Hall of Albany is very hostile towards you & the Smith'n Inst. in consequence of something in regard to Hayden?"[38]

If Hayden knew of Hall's resentment at the time, he was probably too preoccupied with other matters to give it much thought. He had chosen his course, cast his lot with Vaughan and Baird, and was looking to the future. Briefed by Meek, and mindful also of Baird's wishes with respect to "the Zoological departments," Leidy's instructions about fossil vertebrates, Engelmann's botanical needs, and the special interests of Kirtland, Dewey, LeConte, and others for whom he was to collect, Hayden readied himself to be gone from civilization for at least a year in "that wild country," the Upper Missouri. His was to be no ordinary venture, but one that was perhaps unique in the annals of American science: a sort of one-man natural history survey of Nebraska Territory, a region so vast that in time it would be subdivided into five large states. An utterly impossible undertaking? So it might seem, particularly now, as that region continues to receive intensive investigation even after the lapse of more than a hundred years. But Hayden was undaunted. He would, as he had assured Baird, do the best he could; no one could do more. His most cherished hopes were about to be realized, and he was full of high enthusiasm, determination, and vigor. These were attributes he would need, with courage and resourcefulness too, if he were to accomplish his task and return again safely. And he was also to need good luck.

11

To Fort Sarpy on the Yellowstone

Dr. Hayden and his patron, Colonel Alfred J. Vaughan (also spelled Vaughn) left St. Louis on June 1, 1854 as passengers on the steamboat *Genoa,* which was under the command of a master named Throcmorton. Vaughan traveled as Indian agent to the Upper Missouri tribes, in which capacity he was responsible to Colonel Alfred Cumming, the Superintendent of Indian Affairs at St. Louis.[1]

Vaughan was addressed as "Colonel" not because of any military status but by reason of his post in the Indian Service. Indian agents were political appointees who, once they assumed office, automatically became "majors" or "colonels" whether they "wore the epaulets of an officer or the broad-brimmed hat of a Quaker."[2] Perhaps, as Flora Warren Seymour suggested, the rank of "colonel" was reserved for elderly agents, or for those with many years of service. Evidently Vaughan's seniority was such as to create ambiguity with reference to his title; when he and Hayden joined forces he was called both "major" and "colonel," though usually the latter.

Whatever his title, Vaughan was no novice at his work for he had been in and out of the Indian Service through a period of twelve years. This was a noteworthy record, as many Indian agents did not last long in office, their tenure being cut short by political vicissitudes, ineptness, and the difficulties and hazards inherent in their work. Vaughan was an agent who stayed in office long enough, and took his duties with sufficient seriousness, that it might be said of him he had made a career of Indian work. His name frequently appears in the chronicles of the period and the region, and his annual reports to his chief are informative documents. Those records, however, give only glimpses of the side interests which led to his unique connection with Hayden.

Vaughan was a southerner who had grown up in Virginia. He was probably born there in or about 1801. He served for several years in the Virginia legislature as the representative of Dinwiddie County, and it is recorded that while a resident of the Old Dominion he filled various other "positions of honor and trust." Later he moved to St. Joseph, Missouri, and entered the Indian Service in 1842. He became agent at the Osage agency in 1845, and was sub-agent for Iowas, Sacs, and Foxes in 1853-1861. Thereafter he served in the Missouri legislature as the representative of Buchanan County. According to his political confreres, he was a "consistent and energetic Democrat," "an influential and talented member" of the legislative body and one of "the old fashioned State Rights School of Politics." The agency for the Upper Missouri tribes was created in 1852, and when its first incumbent, James H. Norwood, was murdered late that fall, Vaughan sought the post. Backed by members of the Senate and House (more than seventy of whom signed one testimonial in his behalf addressed to President Franklin Pierce) he secured this appointment early in 1853. Thus when Hayden first met him a few months later, he was making his initial tour to the Upper Missouri tribes.

William T. Hamilton, who met Vaughan in 1857, described him as "a fine looking old man" (though he was then but 56).[3] The fur trader Charles Larpenteur characterized Vaughan in terms which have been widely quoted: "a jovial old fellow, who had a very fine paunch for brandy, and, when he could not get brandy, would take almost anything which would make drunk come. He was one who remained most of his time with his Indians, but what accounts for that is the fact that he had a pretty young squaw for a wife; and as he received many favors from the Fur Company, his reports must have been in their favor."[4]

Anne McDonnell considers this "the usual sour comment of Larpenteur" but adds that "no doubt there is a great deal of truth in it." Vaughan's reports to the Commissioner of Indian Affairs, 1853-1861 indicate, she says, that "he spent a great deal of his time with the Indians in his charge, and made long and arduous journeys to contact the various tribes in his district. He had an earnest desire to perform his duty toward the Indians and also had a practical knowledge of what was best for them. He worked hard to carry out the obligations of his office."[5] In general, this estimate is shared by John C. Ewers, who states that "Larpenteur's characterization should be taken with more than one grain of salt. It was his habit to damn virtually all Indian agents and some very able fur traders as drunkards. As for Vaughan's favoring the fur traders, it simply was not true... Actually, Major Vaughan was an experienced and able agent who conscientiously tried to implement the government's policies for the civilization of his Indian charges."[6] This, then, was Hayden's sponsor and closest companion on his second trip into the West.[7] The figure of Colonel Vaughan appears often in the pages that follow.

Hayden's activities in 1854 and 1855 have been reconstructed from his many letters to Baird, occasional letters to Meek, Engelmann, and Leidy, and from Vaughan's official accounts as published in the 1854 *Annual Report of the Commissioner of Indian Affairs*. (Missing are Hayden's journals and other notes, and any letters he may have written to Dr. Newberry, Professor Kirtland, Professor Dewey, other colleagues, or to his relatives.) Some of Hayden's letters were written on the *Genoa*, in which case they were usually headed as coming from the next settlement or fort where the boat would dock and mail could be posted. Other letters were written at these places during intervals between trips. For the most part the letters give only highlights and sketchy accounts of his activities, or his plans for the future, the latter often being uncertain. However much one might wish for the continuity and details of Hayden's lost journals, with their day-by-day narrative of his experiences, the letters do shed much light on his travels, and reading between the lines one can infer many things not explicitly stated. Also, Vaughan's reports, which were prepared with Hayden's help and contain sections actually contributed by him, help to fill in the gaps.

Sometimes Hayden wrote under circumstances which permitted only a hastily scrawled paragraph or two. Again, he would begin a letter and then add to it from time to time, at different places. Under more leisurely conditions he wrote long, informal letters. These are utterly unlike most of his letters written in later years when, in the hopeless attempt to keep abreast of an overwhelming correspondence, he resorted to a terse, telegraphic style of writing devoid of any superfluous words.

When circumstances permitted, Hayden would write several letters, one after the other. Those bearing the same or nearly the same date naturally cover much the same ground, but there is less repetition in them than one might expect; that is, accounts of the same trip or experience, written to different persons, are not in every respect identical. In writing to Baird, Hayden naturally stressed his activities in "the zoological line," which embraced collecting everything from insects, fishes, amphibians, reptiles, small mammals, and embryos, which could be chucked into containers filled with alcohol, to birds and mammals which had to be skinned and stuffed, and very large mammals, like bighorn sheep and bears, from which only the skins and skulls could be saved. To Engelmann he wrote "botanical letters," to Leidy he reported what he had found, as he put it, in the way of "bones," meaning the fossilized remains of mammals and reptiles. To Meek he disclosed his geological findings, wrote about his collections of Tertiary and Cretaceous invertebrate fossils, and touched on many other subjects of mutual interest, especially questions relating to when, where, and how they could investigate the fossils and prepare reports on them. The letters give evidence of his interest in Indians, too, their languages, cultures, and customs, and during the journeys of 1854 and 1855 he may have collected Indian materials, as he did in later years. However, his circle of acquaintances did not include any correspondent who specialized in Indian studies nor is it possible to ascertain exactly the extent to which Vaughan's reports incorporate Hayden's observations and views on Indian affairs.

After departure of the *Genoa* from St. Louis, Hayden's first letter to Baird was written on June 4, 1854 and headed simply "Missouri River." It was a short letter and probably was mailed at Jefferson City or Boonville. In it he wrote:

> The boat is rocking so that I do not know that I can write so that you can read it, but I feel anxious that you should know of my prospects so far. I received a letter from you just before the boat started. I am glad so many people felt an interest and contributed. I will try to satisfy them.

Col. Vaughan… has spent about 40 dollars in outfit, so that I have got about all that was necessary. He will get some more things at St. Joseph's. I got 2 kegs of 5 gallons alcohol, 10 gallons — about 10 pounds arsenic, 6 reams of paper for plants, besides some wrapping paper, a variety of tools, etc. He spent himself something like 200 dollars for provisions and things to pay for help…

I saw Dr. Shumard. There seemed to be no difficulty in letting me off. I found that Prof. Swallow was not permitted to make me any larger offer. So that is all right… I do not wish any one to feel hard toward me but I wish to do what seems for the best, and I could not for the sake of Natural History let this opportunity slip. I suppose government will do nothing up here for many years if ever.

On the whole my prospects are very bright and fair now. I feel there is a contingency about the thing. Vaughan, being a man of a good deal of perseverance, I presume he will carry it through if he lives and I think will not dare use me ill…

Dr. Engelmann let me have $50, which I was obliged to use. I intend to stay as long as possible, perhaps a year & half. This will depend on success and Col. V.'s notions. I feel very anxious to visit the Bad Lands again, and shall endeavor to do so if possible.[8]

A week later, Hayden posted another letter to Baird:

Council Bluffs, June 11th/54.

Prof. Baird,

My Dear Sir,

I write you now for the last time until the boat returns from the Yellow Stone. I shall soon leave the region of post offices…

Thus far all has gone well. The boat started June 1st and now we are within a few miles of Council Bluffs. … At C. B., the boat will discharge a great quantity of freight so that the boat will be comparatively light and will make the trip to Fort Pierre in a very few days.

My prospects are all I can wish. Col. V. is deeply interested in the enterprise. Many things have occurred to render his interest greater still. He now talks of keeping me at work two years but I think I shall get through by a year from next August and send my collections down by the boat on its return next summer. This will give me a chance to explore a rich field and I hope I shall be able to explore it well.

I have already made a pretty good collection of plants. Vegetation is not as forward by two or three weeks as last year. I have found one snake… different from any I have ever seen in the states. When I caught him he flattened himself out like a knife blade, his head assumed the form [of] a spear head. His body is covered with dark and purple spots. He had swallowed an enormous toad which he disgorged as soon as I caught him. I presume he is familiar to you.

I would like to ask a favor of you if it will not be asking too much. I have never been able to get any reports published by Congress though I have written often. Should it be in your power to lay any of these aside for me I will willingly pay you any expense or trouble you may be [put] to.

I will continually keep you advised of my success, good or ill. I shall ever be grateful for the deep interest you have taken in my welfare and I trust I shall yet be able to testify my gratitude to you in some other way than by words.

I shall ever remain with the kindest regards,

Yours,

F. V. Hayden[9]

In his report for 1854, Vaughan states that the *Genoa* reached the village of the Yancton band, near Fort Lookout, on June 18th. Here was assembled the entire band, "who, having been informed of the approach of the boat, made every preparation to receive their agent, and signified their joy by the firing of guns, singing, and other demonstrations." The distributions of goods to this band took place as follows:

> Their village being on the bluffs beyond, a spacious lodge was erected on the bank of the river, and I there distributed to them the portion of the presents intended for the Yancton band. They were received with the most evident indication of satisfaction, and more especially were they pleased with the agricultural implements the government was so kind as to send them. The game has left their country, and they know they must resort to the cultivation of the soil for their subsistence…
>
> Having had more frequent communication with whites than other tribes who inhabit distant districts, many of their chiefs and principal men having visited the States on different occasion, they can realize the superior advantages to be obtained by education. They have seen the comforts attendant on agricultural operations, and are aware of their incapacity in this respect; also… that unless a commencement is made to till the soil, the day is not far distant when actual want will compel them to fight their way into the country of their enemies in quest of game, or remain at the homes of their fathers and perish.[10]

In his report Vaughan urged that the time was now suitable for the establishment of missions, by which to encourage agricultural and pastoral pursuits among the Yanctons, and he asserted that the members of this band would heartily cooperate with their preceptors to obtain this end. He added:

> From what I can judge, and from long experience in the Indian country, I would suggest that, respecting useful teachers, the Catholics appear to be the most capable. It is not from any personal attachment to that sect I am induced to propose them as most competent; neither do I believe theirs is

> the only true religion. I have been taught, and am a believer in, the protestant faith and mode of worship, but candor compels me to say they are the only people who have shown sufficient zeal, knowledge, and industry to carry out any extensive measures the government may entertain towards the civilization of the Indians. Wherever they have placed themselves, they have been uniformly successful; besides, the rites and ceremonies of the Romish church are imposing, and would attract the attention of Indians first; afterwards the realization of its sacred truths would follow. They are, in short, indefatigable laborers in the vineyard, and know precisely the amount of instruction unenlightened minds can bear.[11]

These themes, the urgent need for helping the Indians adjust themselves to new modes of life, particularly agricultural pursuits and the superiority of the Catholic missionary enterprises, are repeatedly stressed in Vaughan's annual reports.

Fort Pierre was reached on June 22nd. Here, reported Vaughan, the "Sioux had been long awaiting the arrival of their agent, and congregated in large numbers on the bank to greet the arrival of their friends." During the day, presents were distributed to several bands, namely the Brulees, Two Kettle, and Sans Arcs, which received them "with satisfaction and gratitude." These bands, said Vaughan, also expressed a desire for peace, and a determination to abide by the stipulations of the treaty at Fort Laramie.[12]

Continuing on its course, the *Genoa* reached the village of the Yanctonnais, 250 miles above Fort Pierre, on June 27th. Here, by prearrangement, the Indians were promptly on hand, "waiting anxiously to receive their presents of this and last year." Vaughan reported that

> the quantity appeared to surprise them, and they manifested great feelings of satisfaction. I held a large council with them, pointed out the disastrous consequences of war and the committing of depredations upon whites. To the agent's requests, the principal men promised strict obedience, stating that they would urge upon their people the destructive character of war and the advantages of peace.

On the following day the boat reached Fort Clark. Here the Arickarees were encamped. These Indians "gathered their whole force on the bank and greeted the arrival of the boat." Vaughan distributed to the Arickarees their portion of the presents; these, he stated, "they were overjoyed to receive, and expressed themselves perfectly satisfied, and grateful to their Great Father for his kindness."

The next morning Vaughan gave out goods to Mandans at their small village about four miles above Fort Clark. Concerning the Mandans he stated in his report: "This nation have always been remarkable for their good disposition towards whites, and are desirous of being at peace with the nations around them; and now they are too much reduced to enable them successfully to contend with their enemies. I found them in a very destitute condition, and the distribution of goods to a nation so needy gave me as much pleasure as themselves."

Later on June 29th, the *Genoa* reached the Gros Ventre village at Fort Berthold. The Gros Ventre, likewise, "had long been waiting with anxiety the arrival of their annual presents, and received the same with the most evident tokens of gratitude."[13] They, like the Arickarees and Mandans, expressed the wish that no more corn be brought them, because they themselves raised enough for their own use and to sell to traders and other nations.

Several letters written by Hayden during the first days of July, for posting at Fort Union, give much additional information about the trip up the Missouri River, and particularly his own pursuits in natural history. The first of these letters was addressed to Meek:

Fort Union, July 1st/54.

Friend Meek,

I date my letter Fort Union though we are as yet in the Big Bend of the Missouri near mouth of White River. But the boat will only stop a few hours before it returns so that I must anticipate somewhat and write on the boat.

I felt a good deal interested to know the result of your decision with Prof. Hall. I suppose, however, that you are still plodding away at your old post bound by a chain. I wish you could do something for yourself. At any rate I hope you will be so [kind] as to lend a helping hand in describing my fossils when I return. I shall have at least a good sized cartload, a great many of which have not been dreamed of in the world of science yet.

I have had a pleasant voyage, a good deal like the [one] we made last season. We have been now just 30 days out and will make Fort Union in 32 I think. A short trip. The American Fur Company's boat did not start for two days after ours and passed us just below Fort Pierre. Both boats lay at Fort Pierre together, a thing that never occurred before, two steamboats lying at one place above Council Bluffs. They will probably make the whole trip in about 29 days. Old Sarpy is head man again.

We had a great time this morning killing a Grizzly Bear. One was crossing the river and they put the shots at it as they did at that poor Elk last summer. Finally one fellow shot him in the head and killed [him]. I ate some of him for dinner today. The Captain is stuffing his skin to put on top of the boat as a trophy.

We have seen lots of bear and today we have seen three bands of buffalo. While I am writing this a large drove are quietly ruminating on the bank a few miles ahead just in sight.

I have made some very fine collections though the boat has made no delay of any kind over 6 hours in length, not even for Indians. I have got 500 species of plants, an open large jar holding two gallons filled with fish, reptiles and mice, etc., a fine lot of bugs and some birds, a prairie dog's skin. About three good sized boxes of fossils. I did not cross the Big Bend below Fort Pierre. We came to it late in the day and camped in it over night, so that

I found no fossils below Fort Pierre excepting Ostrea congesta and fish scales. About 300 miles up the river from Fort Pierre I found on the top of high conical hills, a mile or two from the river, a conglomerate of shells… I took a man with me and we brought all we could. I can load a boat with the finest kind of shells. I have also found any quantity of petrified wood of the finest kind, far superior to any we saw.

I will give you one or two sections which I do not wish you to show to any one, but I have my reasons for thinking that it will reveal some important facts yet. I took a section at Fort Clark, from the water up:

[Geological notes]

I also took a section at Fort Berthold. Dr. Evans says the coal [is] from 4 to 6 feet thick here. I could find it only 2 to 2 1/2 ft. in any place. There is about 2 ft. of dark clay over it which might be mistaken for coal at a distance. I took my pick and clambered and dug in, sometimes two feet, so as to make all sure. From some other things I have seen I think Dr. Evans is not so careful in his observations as he might be, and I think takes some things from heresay. My section at Fort Berthold is somewhat different. The lignite is a higher bed I think. The dip is up the river. In the alluvial banks I find many species of shells… though I have not found a single live shell on the river yet of any kind.

I saw Shumard at St. Louis. He wanted to describe my fossils but I told him you had already spoken to me and I prefered to have you. He treated me very kindly indeed.

I do not know when I shall return, probably not for two years or more. We shall stop at Fort Union until September and then go to the Crow country on Yellow Stone. In the mean time I shall take a trip of 3 or 400 miles down the river on both sides. The country is exceedingly interesting, much broken.

Write me as often as possible, and believe me ever yours sincerely,

F. V. Hayden

Make any suggestions you can and send me anything that may be published on this country.[14]

Other details about the trip are set forth in a letter that Hayden wrote to Baird the next day:

Fort Union, July 2^{d}, 1854.

Professor Baird,

My Dear Sir,

We are now fast approaching Fort Union and as the boat will stop there only a few hours, I must anticipate somewhat… The boat has made an unusually quick trip, not giving me as good a chance to collect as I could wish. Much of the wood was already cut, and no winds or other troubles to detain so that the boat has stopped no where for Indians even, over 6 hours at a time. So my collections are rather small, but I hope of some value... I endeavour to be as minute, accurate and patient in my investigations as possible, so that my conclusions will not hereafter be found baseless.

My general health has been quite good, but I have been grievously afflicted with "sore boils." The first one came on my "seat," of enormous size and made me down sick for two or three days. Since leaving Fort Pierre I have had one on my knee which has very much crippled my efforts as well as myself. I suppose it is the result of change of climate and diet...

I can give you no idea of the interest this country affords to the Naturalist. As soon as I get free from the boat I propose to spend 3 or 4 weeks ranging up and down the river, visiting Forts Clark and Berthold, etc. Whether Col. V. will go to the Crow country or not, he cannot determine until he reaches Fort Union. He has heard that no Crows were there ready to receive their goods. If so, he will not go, much to my regret.

I send you a jar of Reptiles and Fishes, etc., all I have yet collected. If there is anything new or rare, please inform me in your next. Are my labels good for anything?

No. 1.	Frog, near Yankton trading post 250 miles above Plat [*sic*] River, 100 yards from river on upland prairie. June 25th.
No. 2.	Near Big Bend, below Flat [?] River, Iowa. 20th June.
No. 3, 4, 5.	From Upper Cedar Island 50 miles below F. Pierre. June 19th.
No. 6.	Sergeants Bluff, under a wood pile. June 13th.
No. 7.	Mouse, in an old stump, L'Eau qui Court River. June 16th.
No. 8, 9, 10.	Mice, Sergeants Bluff, under wood pile. June 13th.
No. 11.	Wren, near mouth White River, below Fort Pierre. June 18th.
No. 12.	Near White River on upland prairie. June 18th.
No. 13.	Crawling on the top of bushes [?], at landing Fort Pierre, a singularly active snake. June 22nd.
No. 14.	A large adder, running in road, ½ mile east [?] of Fort Pierre in prairie. June 22^{d}.
No. 15, 16, 17.	Toads from Big Cedar Island, 100 miles below Fort Pierre. June 19th.
No. 18, 19.	Near Council Bluffs. Striped snake. June 10th.
No. 20, 21.	Near Fort Pierre, June 22^{d}.
No. 22.	Near St. Joseph's Mo. June 7th.

At the bottom of the jar are three fishes, caught at the mouth of Cannon Ball River, taken from Mo., June 26th; also on top, 2 more from Big Bend, July 1st. I have put in all the insects I have caught between Fort Pierre [?] & Fort Union, also some shells taken from alluvial banks. I have not seen a live shell since I started. I also send a small bottle of bugs collected between Sergeants Bluff and Fort Pierre… Fish are rare in Mo. but in tributaries are plenty. I hope to send you a fair lot in the Fall, from the Yellow Stone & tributaries. The scale fish in jar is called a "Campbellite" up here…

I should like to learn if any appropriations have been made as yet. Please inform me of anything government is going to do in exploration. Col. V. says it has already been decided to send 100,000 dollars worth of things to Blackfeet next year and make a treaty with them. If so that would afford a good opportunity to explore that region. I wish you would tell me about it.

We killed a large grizzly bear yesterday morning in the river… The old hunters tell me there are two kinds. One they call "yellow nosed grizzly," smaller, more savage, longer legged than the other. Is there any thing in this? They seem readily to distinguish the two. It seems intermediate in size between "big grizzly" and the small brown bear. If there is anything in this I will search it out.

We have seen many bands of buffalo, which has offered fine amusement.

Col. V. has written you I believe. You will see by his letter how he feels. I hope what I do while with him will be a small foundation for future effort, if something can be done by the government. A small appropriation of 10,000 dollars would enable me to explore the whole country this side the Rocky Mountains. Can this be done?

Please remember me to all interested friends. And believe me most sincerely and kindly yours.

F. V. Hayden

…Perhaps the specimens sent back better be kept at W. until Col. V. can be there in person, when some division can be made and my part distributed, though let them be examined as soon as you please.

H.[15]

On the following day, Hayden wrote Baird another letter:

Fort Union, July 3d, 1854.

Prof. Baird,

My Dear Sir,

We have just arrived at the Fort and as some things have occurred which I wish to inform you of, and my other letter is sealed, I will write you another short one… A little circumstance occurred this very morning which is very gratifying to me. When Col. V. settled this morning with Capt. Throcmorton he said that in consideration of my industry and attention to my business and my desire to accomplish some thing for myself he would charge nothing for my passage or my freight, and also carry anything back I may wish to send, free. This was also gratifying to Col. V. I have received all the attention and accomodation [*sic*] which anyone could ask and in consideration of Capt. T.'s liberality, Col. V. seems inclined to do still more. So that I cannot see but my joy is complete, my prospect fine…

A keg of specimens will be coming to you by Adams & Co.'s Express. I have done up a bundle of several species of Cacti and sent to Dr. Engelmann. Those birds are of no account. Being on the boat I had no facilities for preparing them. Still I send them, that you may be able to identify them…

I have sent a flour barrel filled with fossils and animals to Council Bluffs, also a box of the same. These are carried without expense.

The Crows are not here. Col. V. will doubtless go up the Yellow Stone. I am very glad of this, though he was in hopes the Crows would be here to receive their goods and save him much trouble.

Hoping you will receive all safe, I remain as ever, sincerely & truly yours,

F. V. Hayden[16]

As noted in Hayden's letter (and also in Vaughan's report), the *Genoa* reached Fort Union at the juncture of the Missouri and Yellowstone Rivers on July 3rd. The unloading took place that same day. Fort Union has been described as "the best built post on the Missouri, with the possible exception of Bent's fort on the Arkansas, the best in the entire West."[17] After visiting this post the year before, Governor Isaac I. Stevens wrote the following about it:

Fort Union… is situated on the eastern bank of the Missouri river, about 2¼ miles above the mouth of the Yellowstone. It was built by the American Fur Company in 1830, and has from that time been the principal supply store or depot of that company. It is framed of pickets of hewn timber, about 16 feet high, and has two bastions, one at the northwest and one at the southeast corner. The front or main entrance is on the side opposite the river. This fort is probably 250 feet square. The main buildings, comprising the residence of the superintendent and the store, are on the front or eastern side. They are two stories high, and built of wood. The shops and dwellings of the blacksmith, the gunsmith, the carpenter, the shoe-maker, the tailor, and others, are of adobe or of wood, and occupy the other sides. These mechanics are mostly French half-breeds, and have half-breed or Indian wives, and many children. There is a grassy plain

Fort Union, built by the American Fur Company in 1830, sketched by a soldier of Gen. Sully, 1864 expedition. (Courtesy of State Historical Society of North Dakota)

around and near the fort, extending to the base of the rising ground, which is a full mile distant on the eastern side. The Assiniboines, the Gros Ventres, the Crows, and other migratory bands of Indians, trade at this fort, exchanging the skins of the buffalo, deer, and other animals, for such commodities as they require. Mr. Culbertson, who has occupied the position of chief agent of the company during the past twenty years, has under his supervision not only Fort Union, but Forts Pierre and Benton also. He is a man of great energy, intelligence, and fidelity, and possesses the entire confidence of the Indians. His wife, a full-blood Indian of the Blood band of the Blackfoot tribe, is also deservedly held in high estimation. Though she appears to have made little or no progress in our language, she has acquired the manners and adapted herself to the usages of the white race with singular facility. Their children have been sent to the States to be educated in our best schools.[18]

Disappointed in his hope of finding the Crows assembled at the mouth of the Yellowstone, Vaughan at once sent runners to the different camps to call them in for the distribution of goods. This measure brought results. According to his report:

In a few days the band Gens des rosches, about eighty lodges, came in company with perhaps thirty lodges of Cree [Cre`e] Indians. To the former I gave the proportion of goods allowed, estimating it per lodge; and after many speeches and other manifestations of friendly feeling towards their Great Father, the soldiers of that band, appointed by me, separated the merchandise among their people to the entire satisfaction of all.

The Cre`e Indians not being a party to the treaty at Laramie, were consequently excluded, although I cannot help but think it would greatly benefit the government, and tend towards forming a general peace, if these Indians were also taken under our protection. They are now the most numerous and inveterate enemies of the Blackfeet, with whom the government are anxious to conclude a treaty of peace; which, as far as regards the Indians of this district, cannot be of long duration if the Cre`es are excluded, because the geographical position the latter occupy is such as to bring them into constant collision with the Blackfeet, who inhabit the same region. They [the Cre`es] are very numerous, good warriors, peaceably disposed towards whites, and come within the American boundary.

Vaughan and Hayden remained at Fort Union for more than two weeks. Vaughan sets forth certain of his activities during this period with considerable detail in his report:

A day or two after the above-named Assinaboines had departed, the rest of the nation arrived and encamped midway between the forts of P. Chouteau & Co. and Harvey & Co., which location the Indians preferred. Invitations were given to Major A. Culbertson and other gentlemen of either fort to attend the distribution, which they accepted, and were present.

On this occasion about three hundred and fifty lodges were present, consisting of the band Gens des Roaches, or Stone Indians; the Gens des filles, or Girls' band; and Les gens des caruts, or Canoe Indians; all of whom, with their families, were encamped in skin tents. The whole were headed by the "Foolish Bear," the same who was appointed chief of the nation by the commissioners at the Laramie treaty. Many other chiefs and soldiers were also present, some of whom ranked higher as war-chiefs than the "Bear," but were more intelligent and tractable than he. Each of these chiefs had their several bands ranged in front of their camp, with the American colors flying, and the whole presenting a lively and interesting scene.

The merchandise having been hauled from the fort to the centre of the camp, the chiefs and soldiers were invited to assemble around that spot, where a series of speeches were made, which are entirely too long to be inserted here, but which deserve to be noticed.

They admit and admire the benevolence of their Great Father in sending them these presents, and feel very thankful for them. They have acted on all occasions in conformity to the spirit of the treaty, and still intend doing so, notwithstanding the continual aggression of other tribes. Their complaints are that some parties of Sioux warriors stole their horses and killed some of their people last winter, and that they [the Assinaboines] were, according to the spirit of the treaty, debarred the privilege of revenge by warring against them. That the Blackfeet, a warlike and hostile nation, were almost daily sweeping off their horses from the doors of their lodges; but that, being aware of the desire of the government to stop war altogether, and of their intention to make a peace with this nation, they have heretofore refrained from any endeavor to retrieve their losses by war in that direction.

Many good and sensible speeches were made; to all of which I gave the general reply, that their Great Father never ceased working for their benefit; that everything could not be accomplished in a day, but that his present policy would continually bring about peace and happiness to all. I pointed out to them the comparative safety and comfort in which they now live to that of former years, before a treaty had been made, or a large amount of goods presented to them. I enforced upon their minds the necessity of submitting to a few present evils, with the view of securing future prosperity. That their Great Father was an indulgent parent, but also one who could and would punish offenders; that his children would always be well taken care of, and the bad become poor and miserable. The result of the whole was, they are convinced that the government pursues a humane policy towards them, and they are confirmed in their resolution to observe the treaty stipulations.

During the whole time, and in all instances, they expressed themselves in very friendly terms towards myself, and appeared to be highly pleased with "talk" directly from their Great Father.

This nation being separated into five different bands, and their hunting-grounds being several hundred miles apart, in conformity to their request, I recognized a sub-chief for the governing of each band, when separated from the others, the whole under the head chief "Foolish Bear." This is a most excellent regulation, for in this way the government secures the friendship of the principal man in each band, who in turn has the soldiers of the camp under him, they forming the only coercive force on the movements of the

> camp.
>
> By this subdivision of authority, it is almost impossible for a war party to leave a camp, as the chief and soldiers of that camp feel bound to protect them. Besides, this advantage of position gives rise to a desire to preserve order, and to repeat on different occasions around their council fires the good advice received from their Great Father through his agent, which cannot fail to have a great bearing on their general conduct.
>
> The goods presented made a considerable display. I dressed several of the soldiers and the chief of each band, who, after forming their forces into separate circles, distributed equally among all; while songs of thanks were sung through the camp by the old sachems. The whole wound up by several dances in P. Chouteau, jr. & Co.'s fort; and in a day or two they all went on their way rejoicing.

Several paragraphs of Vaughan's report are devoted to the special problems confronting the tribes in this remote region:

> On viewing the present condition of the Indians in this district, I cannot but regret their improvidence and indisposition to turn some of their idle time to the purposes of cultivation, instead of relying wholly on the chase for subsistence. Although for many years buffalo have been tolerably numerous throughout the country of the Crees and Assinaboines, yet they are rapidly diminishing, and other nations have but few; besides, their range is gradually becoming more limited; and the extensive plains between the Coteau de Prairie and Saschawaine, formerly covered with these animals, are now entirely deserted by both Indians and game. The Cree and Chippewa tribes, unable to find game on their own lands, are obliged to hunt on the tributaries of the Missouri. The migration of both Indians and buffalo is westward, and the few herds of these animals left are surrounded and killed in the winter on the banks of the Missouri.
>
> The enormous destruction of these animals for their hides, meat, etc., by accidents in crossing rivers on the ice, where thousands sink by becoming imbedded in mud and snow, by storms, and wolves killing the small calves, must, before many years, end in their entire extinction, or at least render them so scarce as to be inadequate for the subsistence of the numerous tribes of Indians who now live by hunting.
>
> In the winter of 1846, the buffalo disappeared from this district, having taken another range on account of the grass on the plains being burned the fall preceding. Most of the Assinaboines were encamped on the Missouri, where they subsisted for a time on elk, deer, and wolves. But these Indians are no deer-hunters; and even if they were, small game is not found in sufficient numbers to support them. The snow was deep, the ground frozen to the depth of four feet; consequently roots, herbs, and berries, their usual resort in times of great scarcity, were not to be found. After eating up their reserves of dried berries and roots, they subsisted on the flesh of their dogs and horses; these failing, actual famine came upon them. As soon as the snow thawed, they separated and scattered through the whole district, even to the banks

of the Saschawaine and Red rivers. Some were so fortunate as to find a stray bull or antelope — others nothing. Many died of disease and hunger; old persons were left behind to perish, and in several instances they ate their own children.

This fact is mentioned to show what misery would assuredly follow, were the buffalo their only reliance, driven from their country by emigration through it, or in any other manner reduced so as to be insufficient to supply their wants. It is therefore due to humanity, to our national honor, as a free, rich, and enlightened people, that some foundation should be laid in time for the future welfare of the red man.

It is certainly discouraging to commence agricultural operations among people whose confirmed habits are at direct variance with such pursuits; but were a mission formed among them on the principle of manual-labor schools, the young could be brought up in industrious habits and knowledge, which many of the grown Indians could be induced to realize the benefit of, and pastoral employments joined with a certain amount of agricultural labor. To effect this, the Indians, or at least a portion of them, must become stationary; the Indian agent reside with them constantly; war be entirely stopped by treaties or otherwise; good teachers, farmers, and mechanics employed, and suitable amounts appropriated to meet these expenses. This is at least worthy of thought, if not of trial.

These Indians, as before stated, have strictly observed the stipulations of the Laramie treaty, and it is gratifying to know that not a single instance of murder, robbery, or other depredation, has been committed by them, either on the neighboring tribes parties to the treaty, or on whites. This is the more remarkable, as before the treaty they were foremost in the van of thieves and robbers, always at war, pillaging whoever they met, and annoying their own traders in their own forts.

In the event of government wishing to open a way for emigration through this district, or the construction of a railroad, I have every reason to believe that, by pursuing the proper policy, these tribes would offer no formidable obstacle to the accomplishment of these ends.[19]

Although Hayden's letters indicate that, as the *Genoa* proceeded up the Missouri River, he seized every possible opportunity to leave the boat in order to collect specimens and otherwise advance his natural history studies, nevertheless he may have been on hand for most if not all of the colorful assemblies described in Vaughan's reports, when government goods were distributed to the Upper Missouri tribes. Indeed, he may have been more than simply an interested on-looker at these spectacles, as it is probable that Vaughan called on him for assistance in the execution of his duties. There is little specific information about this, either in Hayden's letters or in Vaughan's report. It is certain, however, that Hayden's close association with Vaughan on the journey to the Upper Missouri tribes in 1854, and again in 1855, gave him considerable first-hand acquaintance with these Indians and their problems and left him with a real and continuing interest in them. This may be inferred from several circumstances, particularly the vigorous though unsuccessful effort that Hayden made to succeed Vaughan as Indian agent when the latter retired in 1862.

After posting a letter to Baird at Fort Union on July 3rd, Hayden, so far as is known, wrote no more letters until more than three months later. The conspicuous break in his correspondence resulted from a journey which he took with Vaughan and which took them completely out of touch with civilization, a journey notable alike for its strenuous character, hazards, and from Hayden's point of view, exciting rewards. Setting out from Fort Union on July 18th, Vaughan and Hayden with their retinue ascended the Yellowstone River in order to meet with the Crows at the post of the American Fur Company called Fort Sarpy, which stood on the right bank of the Yellowstone about twenty-five miles below the mouth of the Bighorn.[20] When Vaughan's duties here were eventually concluded, the party returned to Fort Union by the same river route. Subsequently they descended the Missouri to Fort Pierre, but evidently not at the same time; Hayden seems to have preceded Vaughan in making this part of the return trip (concerning which little is known). At Fort Pierre, Hayden again took to the river, this time entirely alone, by skiff, in order to bring express down to the nearest settlements. Not until October 6th, when he had reached Council Bluffs, was he able to resume his correspondence. A letter of that date and several others written not long afterward, together with Vaughan's published report for the year, tell what is known about this three-months' sojourn in the wilderness. The story as pieced together from these sources is by no means complete. Nevertheless, it is clear from the available information that the long "trip to the mountains," Hayden's first deep penetration into Indian country, provided him with fruitful and stimulating experiences.

The following account of the trip to Fort Sarpy and return is quoted from the first of the several letters Hayden wrote when he got back to "the region of post offices":

> Council Bluffs, Oct. 6th, 1854.
>
> Friend Meek,
>
> You may think it rather strange to receive a letter from me dated at this place. Well, I will tell you the reason in due time. I believe I wrote you from Fort Union by the return of the boat? After remaining at Fort Union 17 days, during which time I explored the country pretty faithfully for 10 miles around and made many interesting discoveries, we started our trip up the Yellow Stone. The goods for the Crows were all deposited in a keel boat and taken up by cordelle, so that the boat did not travel only from 6 to 12 miles per day, so that I had ample time to search the Yellow Stone pretty thoroughly, which I did for about 5 miles each side from the river.
>
> You will see by my collection that I was not idle, although I did not save [more than] a little of what I collected, for if I had a steamboat [it] would not have contained them [all]. As it is I have about two tons and [a] half of fossils, several gallons of alcoholic specimens, as bugs, fish, snakes, etc. About 12 or 15 hundred spec's plants, rather more than last year. A good collection of recent shells... some which I think are new. Of animals 4 fine beaver well stuffed, 2 grizzly bears, 8 prairie dogs, several [?] ducks and other birds, foxes,

wolves, a half dozen fine deer, antelope, big horn heads, stuffed with the skin, you know, about 6 inches of the neck, making rather beautiful specimens. At any rate the collection fills a good sized room at Fort Pierre, all [of] which has not cost the Col. a cent yet, so that if it will not bring much there will not be much lost.

I cannot describe within the pages of a letter or even give you any idea of that wonderful country the Yellow Stone. My notes are at Fort Pierre, very copious, taken down on the spot with as much accuracy as possible, with spec's to illustrate sections. I have a section of every bluff that makes its appearance up the river and specimens to illustrate the geology to the mountains which a friend of mine obtained for me. So I have a pretty fair section from the mountains to Council Bluffs.

One thing may surprise you. I have never in the Carboniferous formation seen a finer exhibition of the characteristic [?] flora, a few ferns but mostly dicotyledonous plants, most beautiful and perfect. The Lignite Beds I have examined pretty well, and have a good set of fossils to illustrate them. Great quantities of splendid petrified wood, etc.

Although the collection might have been more perfect and larger, yet I could do no more single-handed as I was. I was obliged to take my pick in one hand, my bag in the other, my note book in my side pocket, my bottle of alcohol in my vest pocket, my plant case around my neck, and with all these carry a gun to defend myself from bears and Indians. So you see this was studying Natural History under quite a burden if not under a "difficulty"... I have Cretaceous fossils that will do your eyes good to look at them...

[On the return journey] we started from Fort Sarpy on the Yellow Stone 18th Sept., stayed at Fort Union 4 days, 18 days coming from Fort U. to Fort Pierre. There was no one there to take the express so I started in a little skiff. 2 men have just got [here] where I shall spend a week or two, hunting around in Nebraska, and then go up to Fort Pierre on horse. Whether I shall stay in the mountains another year I cannot tell. Col. Vaughan & Culbertson are anxious to have [me] follow the cordelles up to the Blackfoot next Spring. But it [is] getting very dangerous indeed. In fact I fear I shall not be able to go to Bad Lands next year...

Ever yours,

F. V. Hayden[21]

Writing to Meek again, from Bellevue on November 11th, Hayden added a few additional details about his trip:

Coming down from Fort Pierre... though in a great hurry with express I made some notes of the Geology, which added to yours I think will be of some value. The river was extremely low and I found some beds that we did not see last summer. I also found several new species of fossils, a fine coral at Dixon's Bluff, etc. At the mouth of White River I found some interesting bones, at lower end of Big Bend left bank I found a heap of chalk fossils, one or two new species I think. I have been spending some days here collecting Carboniferous fossils. Got some fine ones…[22]

A letter to Baird, also written on November 11th, reviews Hayden's trip to Fort Sarpy and return:

Bellevue, Nebraska, Nov. 11th, 1854.

Prof. Baird,

My Dear Sir,

I was very sorry that I did not receive a letter from you of later date than 3d June, at the time of the fall express. Perhaps they were lost or mislaid. Col. Vaughan wrote you from Fort Union about 1st of July. I was in hopes you would give him an encouraging reply. He has written to you again this fall. Please give him a little touch on the importance of scientific researches. It will make him more liberal and enthusiastic.

You may inquire on seeing the place & date of this letter how I came here. After returning from the Yellow Stone to Fort Pierre there was no one to whom Col. Vaughan wished to entrust his express. The excited state of the Indians rendered it unsafe for me to operate around Fort Pierre this fall so I consented to come down with the express. I had a hard time of it, a small skiff, heavy winds. But I accomplished what I wished and that was to complete my geological section from the mountains to Council Bluffs…

We stayed at Fort Union about three weeks during which time I rambled about the country. Made some fine collections of fossils, some alcoholic specimens, etc. On the 19th July, all things were ready for our trip up the Yellow Stone… [The keelboat] moved slowly, and gave me a fine time to explore, which I tried to improve to the utmost of my ability. I used to wander off alone for miles from the river… This region, I suppose you are aware, is the debatable ground of all the warlike Indians on Mo. But for the Naturalist, especially the geologist, this is a most interesting region. I could always return loaded down with rare and interesting things…

We had no trouble with Indians until reaching the mouth of Tongue River. There a party… gave us a fight, killing 2 Indians that were with us. The particulars of it I have written in Col. V.'s Report.

It was so dangerous about the Fort on the Yellow Stone that I did not do much there, but I have some alcoholic specimens which I send you. They are not well labeled from the fact of an accident which occurred 4th day out. A barrel containing a fine lot of bottles, my collection, and everything I had ready, stood on the stern of the boat, apparently a safe place. The clerk for the Crow Post in passing around the stern by some means fell into the river and in falling caught my barrel and drew every thing into the river. The young man was drowned.

This accident hindered me from doing as I would have done. I filled all my bottles, also that copper can which is at Fort Pierre. The three bottles I send by stage as no steamboat comes up to this point now. I hope [they] will yet [be] safe, for there are some rare specimens. 2 bottles labeled Yellow Stone came from that [river] and its tributaries. The reptiles came from the high bluffs, fish [from] small streams… & one bottle collected near Fort Union. I shall endeavor to induce Col. V. to make all the alcoholic specimens a present to you — at any rate half of them belong to me you are aware. Please divide them in the Spring when Col. V. comes. Take my portion to yourself and distribute as you think best. So do with all alcoholic specimens I can obtain…

I [will] spend the winter at Fort Pierre. From that point I will write you more fully. I have a great many notes on the birds and animals of this country which I wish to send you. I shall spend the winter among my specimens of which I have a good sized room full at Fort Pierre…

What I shall do next year I know not. The Indians are exasperated. I can assure you that what I do hereafter will be at the peril of my life. I have worked all summer in the midst of danger and I fear it will be worse. What shall I do? I am told it will be sure death to go to [the] Bad Lands but if I stay up there I must go. That is the great center of my operations. I dislike to give it up…

I am in great haste, and pardon my careless writing… Anything in which you can advise me please tell me frankly. It will be for my good and success. I cannot do things with as much system and thoroughness as I could if I had government patronage. But I will do the best I can. I have to work single handed and alone. Col. V. has not yet hired me a man…

Most Sincerely Yours,

F. V. Hayden[23]

Hayden, although a prolific correspondent, seldom took time or felt inclined to dwell at length on his personal experiences except insofar as these related to his scientific pursuits. He was too deeply engrossed in his multiple interests to be greatly concerned about adventures for their own sake. What he had been able to accomplish or hoped yet to do for Science, that was paramount. In the long, detailed letter he wrote Meek on October 6th he did not even mention the Indian attack at the mouth of the Tongue River, and in writing to Baird on November 11th he made only casual reference to this incident. His letters tell virtually nothing about his descent of the Missouri from Fort Union to Fort Pierre. Concerning his trip from Fort Pierre to Council Bluffs with the express, his only comment to Baird was, "I had a hard time of it, a small skiff, heavy winds," yet this was a 400-mile boat trip down the Missouri River, alone through Indian country, in the inclement weather of late fall. As for the return trip to Fort Pierre, he mentioned only that he would go by horse. Hayden's brevity in such instances is exasperating to the biographer and historian.

However, it is evident that Hayden could write an effective story if called on to do so. This is demonstrated by Vaughan's report, which Hayden wrote in part and all of which he may have extensively reworked. As Hayden told Meek on March 25, 1855, "You will see in the Indian Reports, Vaughan's report. I wrote out that report… It is doubtless published ere this."[24] It is not surprising that Vaughan called on his well-educated young companion for such help as this; examples of

Vaughan's letters which have been published make it clear that writing was scarcely the Colonel's forte.[25] In the report for 1854 are a number of paragraphs which, whether written by Hayden or Vaughan or the two of them working together, throw considerable light on what happened during the journey to Fort Sarpy:

> On Tuesday, the 18th of July, the goods for the Crow nation were safely stored in a keel-boat; and all things being ready for our departure, we left Fort Union for a long and perilous trip up the Yellowstone. The boat was seventy-five feet long, loaded almost to the water's edge with the government goods, and those of P. Chouteau, jr. & Co., for their trading-post on that river. This boat was to be taken a distance of three hundred miles, through a most dangerous country, and against an almost resistless current, by human strength, with the cordelle.
>
> We encountered many difficulties; and although we overcame them all, and arrived safely at our destination, the anxiety and almost hourly expectation of having the boat sunk or dashed in pieces on the rocks was the source of the greatest annoyance, and marred entirely the pleasure of our journey.
>
> The descent of the Yellowstone is probably greater than any other river in this country, rolling over a rocky bed, with a current of six or ten miles per hour, filled with sand-bars, islands, and other obstructions. But the most serious difficulty we met with was the rapids and ripples, as they are called in this country, where the current seems dammed up with ledges of rock, over which the impetuous current pours with great velocity. Many of these rapids are as formidable as the falls of the Ohio, at Louisville. At these rapids it required all our strength, men, women, and children. Often the men would pull for hours in water up to their waists, with a current against them which would carry them off their feet at the slightest misstep, making not more than six miles in twelve hours, when it was necessary to be in a continual state of readiness for an attack, not knowing at what instant an unseen enemy might fire upon us from their thousand lurking-places.[26]

The laborious ascent of the Yellowstone in the manner described took almost a month; and when the destination was finally reached, Vaughan was again disappointed to find that no Indians were on hand to receive the government goods. According to his report:

> On the 15th of August the goods were all landed at Fort Sarpy, on the Yellowstone. I immediately despatched three men to the mountains, two hundred and fifty miles distant, in search of the Crow Indians. They returned, after a most perilous trip of twenty-seven days, bringing with them the chiefs and principal men of that nation. These I assembled together in the fort, and held a long council with them, in which I explained fully the wishes and desires of their Great Father, and endeavored to correct any erroneous impressions they might have received respecting the treaty at Laramie, and the designs of the government in sending them these presents. I told them their Great Father did not wish their country in return, (an idea many of them entertain;) that he did not ask of them one spire of grass for the whole; but

> that he only wished his red children might live in peace with the surrounding nations. They asked what they should do when their enemies attacked them, as they are constantly doing in their own country. I told them explicitly, that if their enemies, the Blackfeet or Sioux, came into their country to war upon them, to defend themselves and destroy as many of their enemies as possible; that their Great Father did not expect them to sit quietly in their wigwams and be shot like so many buffaloes. With this they seemed perfectly satisfied, and expressed a desire to be at peace, but said it was impossible under the existing state of things. The Crows have ever adhered strictly to the stipulations of the Laramie treaty, and express an earnest desire to obey the requirements of the government.
>
> The presents were received with marked expressions of gratitude, and the quantity so surprised them that, as they said, they could hardly believe what their own eyes had seen.[27]

The Indian attack which Hayden mentioned to Baird is stated, in Vaughan's report, to have taken place at the mouth of the Tereque River. The word "Tereque" probably resulted from someone's attempt to decipher the word "Tongue" in the scrawl of either Hayden or Vaughan. The date of the attack is given as August 19th, and this also must be an error, as Vaughan's party then seems to have been at Fort Sarpy (they arrived at the fort on August 15th, and departed from there to begin the return journey on September 18th). Perhaps the correct date was August 12th or 13th (that is, a few days before they reached Fort Sarpy). The "August 19th" mistake presumably crept in when the report was set up in type or transcribed from the original manuscript. Hayden wrote Baird, "The particulars of it [the attack] I have written in Col. V.'s Report." No hint is given as to why Hayden, rather than Vaughan, reported the incident. One may conjecture that Vaughan might have been away from the main party when the attack took place, or that he was less advantageously situated than Hayden to observe what happened. More likely, perhaps, is an explanation already suggested: that Hayden aided Vaughan with report work simply because the latter found writing to be a difficult chore. Whatever the reason, the "particulars" provide a graphic narrative, and one that testifies to the dangers which beset these travelers daily, even though they were fully alert, armed, and moved in a body. Yet it was of the Yellowstone trip that Hayden remarked to Baird, "I used to wander off alone for miles from the river... I could always return loaded down with rare and interesting things." The account of the attack, as set forth in Vaughan's report, follows:

> We met with no war parties of Indians until we reached the mouth of the Tereque [*sic*] river, although evident signs of them had been seen some days before; but on the morning of the 19th of August, as we were passing along near the mouth of Tereque river, the Crow Indians who had accompanied us by land from Fort Union, six in number, were riding in advance of us, when suddenly a gun was fired from a thicket of willows, and our Indians uttered a terrible shriek and ran. Fifty shots followed in quick succession, and, with whoop and yell, over seventy naked warriors of the Blackfeet tribe rushed out

> of the bushes to complete the work they had commenced. We were within a few paces of the place of attack, and immediately calling the men on board, crossed over to the opposite side. After completing their work of death, they descended the river about a mile and swam over to the other side. We made all preparation, expecting an attack. As soon as we saw the Indians were all over, we recrossed our boat, and landing near the point of attack, hastily threw up a defence of logs and bush, and then turned our attention to the dead. Two of the Crows were killed, scalped, and their bodies horribly mangled, their skulls knocked in, and otherwise dreadfully disfigured.
>
> After they had done all the damage they could, they ascended the highest butte in the vicinity, and there arrayed themselves as conspicuously as possible, reflecting the light of the sun from their looking-glasses, and by every act and gesture bade defiance to our party. Some of them would come close to us, and, swinging the scalps in the air in sight of the relatives of the deceased, would mock their sorrow. They also killed two fine horses and two colts. We have since learned that this party of Blackfeet, seventy-one in number, started from Fort Union on our trail, having heard there that a boat had gone to the Crow country with goods for the Crows, and that a small party of Indians of this nation were along.
>
> Scarcely a day passes but the Crow country is infested with more or less parties of Blackfeet, who murder indiscriminately anything that comes within their reach. At Fort Sarpy so great is the danger that no one ventures even a few yards from his own door without company and being well armed.[28]

On their return to Fort Union, Vaughan worked on his report (probably with Hayden's assistance) and in it he urged administrative changes which might improve the government's handling of the difficult Indian situation in the Upper Missouri region. He wrote:

> I would most respectfully suggest to the department the importance of establishing two agencies on the upper Missouri, one at Fort Pierre and the other at Fort Union. So great is the extent of country over which the agent is obliged to travel, so arduous the labors he must perform in the distribution of the annuities to the different bands, that it is beyond the power of possibility he can do justice to the government, the Indians, or himself. I would therefore suggest that the Pancaws, Yanctons, Brulees, Two Kettle, Sans Arcs, Minnecowzues, Yanctorinans, Honepapas, and Blackfeet band of Sioux, come under the supervision of the agent at Fort Pierre; and the Arickarees, Mandans, Gros-Ventres, Assinaboines, Crees, and Crows, come within the jurisdiction of the agent established at Fort Union. Such an arrangement would enable the government officers to remain a portion of the time with each band, and give general satisfaction.
>
> All the bands of Sioux have already received their presents with great appearance of friendship, excepting the Minnecowzues, Blackfeet, and Honepapas. The former band are daily expected at the fort, and will gladly receive their annuities; but the Blackfeet and Honepapas still persist in refusing to receive any annuities, and are constantly violating all the stipulations of the treaty. They are continually warring and committing depredations on whites

and neighboring tribes, killing men and stealing horses. They even defy the Great Father the President, and declare their intention to murder indiscriminately all that come within their reach. They of all Indians are now the most dreaded on the Missouri. Something must be done, as the lives of your agents, and of all others in this country, are daily placed in jeopardy.[29]

Vaughan states that, having written the foregoing paragraphs at Fort Union, it was his intention to dispatch his report from that place "to the department," presumably by courier. However, before carrying out this plan he changed his mind and concluded to go on himself as far as Fort Pierre. Reaching this post on the evening of October 18th, he learned of "the unfortunate massacre of Lieut. Grattan and party" near Fort Laramie. Taking particular pains to obtain the facts about this incident, latest of the many bloody conflicts with Indians on the frontier, he extended his report by adding to it, at Fort Pierre, the following "true statement of the affair," one which he thought could be relied upon:

The Wahza-ahe, Brulee, and a part of the Ogal-al-lah bands of Sioux Indians, were all encamped about six miles below Fort Laramie. An emigrant train in passing had one of their cows killed by them. Complaint was made to the commanding officer at Fort Laramie. Some Indians visiting the fort told the commandant that the offender would be surrendered by the head chief if sent for. He immediately despatched Lieut. Grattan with a command of 25 or 30 men; they arrived at the lodge of the head chief, where the lieutenant demanded the surrender of the Indian who had killed the cow. This was refused, and they defied the officer to take him; and from the number of armed Indians crowding around him, Lieut. Grattan saw that their intentions were hostile. He ordered his men to fire upon them, which they did, wounding the head chief mortally, (since dead,) and another Indian. They then discharged the two cannon without effect, having been too much elevated; and before they could reload, the Indians, who awaited ready, closed upon them and massacred the whole party. As soon as this was accomplished, they proceeded to the trading-house of James Bordeaux, and robbed him of everything; and it was with difficulty that the whites escaped with their lives. From thence, they proceeded to the trading establishment of Messrs. P. Chouteau, jr., & Co., where the goods intended to be distributed to them from the government were stored. They broke open the house and took the whole of them, then broke the store of Messrs. Chouteau & Co. open, robbed it of a large stock of goods left from last year's trade, and an entire new stock just arrived from the Missouri, amounting to about twelve or fifteen hundred dollars.

The commander of Fort Laramie was unable to render any assistance or protection to the whites on the Platte, not having men enough to protect his fort had the Indians attacked it, as they at one time stated they would do.

On the following day the whole of these Indians moved off. They are now scattered in small bands within 60 to 150 miles of this place, and are trying to enlist the Indians in this section of the country to join them in a general war on all whites. They keep war parties continually on the road between this [place] and the Platte, and any white man found on the road will certainly be killed by them. They state openly that next spring they will keep parties constantly on the emigrant roads, and kill all they find.

I assure you that my situation here, as well as that of all the traders and their men, at present is perilous in the extreme. There is but one course for the government to adopt, and that must be done promptly — to send a sufficient number of troops of the proper kind, and pursue the bands of Indians who have been concerned in this affair, and chastise them in such a manner that they will not only respect, but fear the government in future. Without a salutary lesson, there is no knowing to what extent they will commit murder and depredations on the whites.

On my way to this place a few days since I arrived at the Yanctonnais village of Sioux Indians, encamped on the river about 100 miles above this place. They generally belonged to a band who have always been more or less averse to carrying out the provisions of the "Laramie treaty." I immediately called a council with the chiefs and braves, and after talking over all matters with them, and giving them good and proper advice, which I thought they received in the spirit I intended they should, I made them a present of some tobacco, provisions, etc.; when, to my surprise, the principal chief, called "Red Leaf," arose with the knife in his hand, cut all the bags containing the provisions to pieces, scattering their contents on the prairie, then threw the keg of gunpowder in the river, and about fifty of them discharged their guns at it. This band of Indians have been very refractory for a number of years; the different agents, who have seen them, have all told them that troops would be sent to chastise them if they did not alter their conduct. They regard this now as only a threat; and until the presence of troops intimidates them they will always be dangerous and troublesome.

At Fort Clark I found one hundred lodges of Aukpapas, Blackfeet, and Sioux Indians, who had been waiting there a month to see me. In the council I held with them the principal chiefs told me that they would not receive any presents from the government, and to do what I pleased with their portion of goods, for they did not want them; that they preferred the liberty to take scalps, and commit whatever depredations they pleased, in preference to goods from their Great Father. They talked very hostile; and I replied to them, that just as certain as the sun rose and set, their Great Father would not only chastise them for their bad conduct, but for the indignity offered to him in not receiving their presents.[30]

In a letter written at Fort Pierre on March 25, 1855, Hayden confided to Meek, relative to Vaughan's annual report for 1854, "like a cussed fool [I] gave him my discoveries on the Yellow Stone. It is doubtless published ere this."[31]

The part of Vaughan's report containing the "discoveries on the Yellow Stone" is easily identified. The report includes several paragraphs which are strictly geographical and geological in subject matter, and only by implication relate to Indians and Indian problems; they form a section of the report which seems out of context, as if inserted in the main body of the text. Though the report as a whole is signed by Vaughan, and phrased as if entirely written by him (Hayden is not mentioned anywhere in the document), this "geological section" gives unmistakable evidence of having come from Hayden's hand and of having undergone little if any change from the form in which he wrote it. It must, accordingly, embody some of Hayden's early observations and impressions of the region bordering the Yellowstone and Missouri Rivers. As his first published account about this region, a part of the vast western territory to the investigation of which he was to give more than three decades of his life, it is worth quoting in full:

> Of the country bordering on the Yellowstone, I would say that there can be none more barren in the world; none of it could be made subservient to agriculture, or even grazing purposes. The land is even more sterile than on the upper Missouri; the timber is very scarce and of diminutive growth, occurring only along the streams and bottom prairies, which along the Missouri are, many of them, quite fertile — here, yield nothing but wild sage, fremontia vermicularis, and the annoying prickly pear. The hills are very barren, sometimes for miles producing no vegetation but a few bushes and stunted cedars. The mineral productions of the Yellowstone are not those that may be said to possess any economical value. Large beds of a poor kind of coal, called lignite, are found along the rivers in great numbers, from one to six feet in thickness, with strata of yellow and white indurated clay intermixed. Some of these beds contain immense quantities of fossil shells, and in the clay portions are found many concretions filled with the most beautiful impressions of leaves and fern. In no country is there a finer exhibition of the ancient flora. The chalk formation occurs only in one locality, where, for the distance of about ten miles, shells were found in the greatest abundance and of the most beautiful forms and hues — probably one of the best localities for shells of the chalk formation in the world. Another remarkable feature of the country, and one which often excited my attention, is the vast quantities of petrified wood scattered through it in great profusion. In some places it is so abundant, and of such magnitude are the logs and stumps, that these regions are called by the traders the "petrified forests." This wood is so perfect that the grain may be seen distinctly, and on the stumps layers of growth as though just cut with the axe.
>
> Another point which must be of interest to the geologist, and which excited my wonder, was, the vast quantities of pebbles scattered along the rivers. In many places where the current has reached, whole acres are covered with them to the depth of several feet, and the whole river is paved with them. But the most remarkable feature that I observed was the numerous appearances of volcanic action that were everywhere presented along the river from the mouth to the source. The distant bluffs present a burnt and reddened appearance, covered with fragments of rocks of every size, in a melted or semi-melted state, which in many places has given the country the appearance of the ruins of ancient cities. This has usually been attributed to volcanic action, but I find, after a minute examination personally, that it is caused by the burning out of the lignite beds, some of which are even now on fire.
>
> This country abounds very plentifully with the various kinds of game common in the west: buffalo, elk, deer, two species of antelope, and wolves and foxes without number. The plains are covered with a variety of birds and fowls, prairie-hens, and a beautiful bird with the habits and nearly the size

of our domestic turkey. It is called the sage-cock, and is, I believe, peculiar to the mountainous regions alone, and might, I think, be domesticated. The streams abound with beaver also, which of late years, from the fact of not being trapped, as formerly, are rapidly increasing. The river swarms with fish and soft-shelled terrapins, furnishing the epicure with many a savory dish. The variety is not great; two kinds of cat-fish, pickerel, a sort of carp or sucker, and a fish somewhat resembling a shad, are the principal varieties. I have nowhere seen cat-fish more abundant and of so rich a flavor. I have myself caught, with a hook, twenty or thirty pounds in as many minutes.

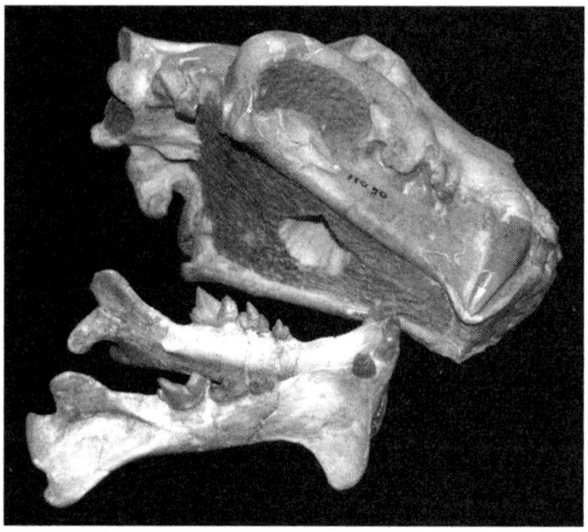

Skull and jaw bone of Sabre-tooth cat; six inches long. (Fryxell Geology Museum, Augustana College; photograph by Erwin Weber)

In speaking of the physical features of this country, that remarkable region called by the traders the "Mauvais Terres," or "bad lands," should not be ommitted [*sic*] in this report. It extends from the forks of the Platte westwardly to the Black Hills, covering an area of ninety miles in length and fifty in width, presenting some of the grandest and most wonderful scenery in the world. Here are to be found scattered over the surface in great profusion the bones and skulls of extinct species of animals, in the most perfect state of petrifaction; terrapins in a fossil state in great numbers, and some of them of enormous size, weighing from 1,500 to 2,000 pounds. Of these wonderful curiosities I shall spare no pains to procure specimens when my duties will permit me, and exhibit them to the department.[32]

While Hayden was on the three-months' journey to Fort Sarpy and back, he must have thought often about his friend F. B. Meek and wondered how he fared. No letters could reach Hayden during this period, however; and he remained unenlightened until early October, when he got back to Fort Pierre. Here he received a long, news-filled letter from Meek. In it Meek told of his altered and much improved circumstances and he sketched his summer's activities. The letter brings us up to date on Meek and at the same time it gives an interesting picture of some of the geological problems of the period, including those which attended field investigations even in well-settled regions like Missouri, where Meek had been at work:

Columbia, Mo., Aug. 13th, 1854.

My dear Sir,

Yours of the 1st July only reached me some two or three days since, after a journey to Albany and back to this place. I was much pleased to hear that you are getting along safely and that you are finding so many interesting fossils and other objects of Nat. History. I hope you will bring back a collection that will add much valuable information to our stock of knowledge of the

Geology and Natural History of that remote and little known region. I was very much interested with your section at Fort Clark. From your description I would suppose this entire section to be made up of some of the recent Tertiary beds or perhaps the higher ones may be of the age of the Loess. These however, are questions that can only be solved by a careful study of their organic remains. It would be well to preserve some good specimens of the shales from the bed above No. 4. by dipping them in thin hot glue. When they are too much decayed to bear handling otherwise, lumps or masses of the matrix in which they are enveloped may be carefully dried and dipped in thin (hot) glue water. In the same way specimens of leaves and other vegetable remains may be preserved when the matrix is very friable. The locality you mention 300 miles above Fort Pierre will doubtless afford quite a number of new and highly interesting species. Of course you will make extensive collections at all such locations.

As you may see by the heading of this letter, I am at last in Missouri. Prof. Swallow offered me a salary of one thousand dollars per annum, and Prof. Hall at last had the generosity to allow me to accept the appointment on conditions that I shall return to Albany about the first of Jany. each year and remain with him until May when I am again allowed to come back to Missouri. Hall is to pay me whatever may be deducted from my salary here in consequence of the four months' absence. Since I arrived here Prof. Swallow has agreed to allow me one hundred dollars per month and my travelling expenses during my stay here. I am to go into a separate district and am allowed to report the result of my own investigations under my own name. Dr. Shumard has some prospects of a more desirable position than that he now holds in our Survey. In case he leaves, Prof. Swallow has promised me his place at a salary of fifteen hundred dollars per annum.

I have not yet been in the field, and will not perhaps start out until about the 1st of September. I arrived here about the 15th of July and joined Prof. Swallow at Boonville. He had left instructions for me at St. Louis to ascend the Osage River to the head of Steam Boat navigation and make a section to its mouth, and thence to the mouth of the Missouri. It was so late however, by the time I reached St. Louis that the Osage was very low and Steam Boats were not moving any distance up it, so that Prof. S. had countermanded his orders and instructed me to join him at Boonville. It was his plan to let me take charge of his part of the Corps and explore some of the counties during July and Aug. while he would be at work at his report. By the time I arrived at Boonville however the Cholera was raging to an alarming extent along the River and there was much other sickness all over the country. This together with the unprecedented hotness of the weather, and sickness in the Corps, induced Prof. S. to allow us all to return to this place, where we have since remained. As Prof. S. is at work on his report and I have been hard at work amongst the fossils and other specimens our time has not been lost to the State. It will be impossible to do anything in the field now until it rains, as all the small streams and most of the best springs are completely dried up so that it would be impossible to find water for our horses in traversing the country. We have had almost no rain for more than two months, and many of the farmers who thought they had never failing springs are compelled to drive their stock eight or ten miles to water. Everything is completely parched up. More than three weeks since the meadow grass was so dry that where accidentally set on fire it burned like prairie grass late in the season. The corn and other late crops are of course hopeless.

Dr. Shumard arrived here a few days back, and we have been at work together on the fossils of which there is already a very fine collection. For about a week we have been debating a doubtful point in regard to the age of a series of rocks from which there is a fine collection of fossils. The difficulty is, that the few species that have been described are mostly such as are common to the Lower Carboniferous and the Devonian formations, so that we are left to draw our conclusions from the affinities of the numerous undescribed species. Shumard is making a catalogue, and we are comparing every species with those considered characteristic of each of the Systems. Sometimes the weight of evidence seems to be on the side of the Devonian & at others of the Carboniferous System. Shumard rather inclines to the Carboniferous while I lean a little to the Devonian. Prof. Swallow also thinks the weight of evidence is on the Devonian side. My opinion is mainly based upon the general resemblance of the fossils to those of the Hamilton & Chemung groups of New York. A few of the species are also identical with Hamilton & Chemung species. For my own part I would be very far from expressing any very decided opinion in a written report, in many cases as to the exact limits either upwards or downwards, of the Devonian System when it is still a matter of dispute amongst some of the best and most profound Palaeontologists of the age whether or not such limits exist in nature, or could be drawn upon Palaeontological grounds.

I anticipate a great treat in looking over your collection when you come back. Of course you will have many new species, and if you do not change your mind in regard to the matter, I would be delighted to join you in investigating and publishing them. I think I have things so arranged now, that I could have ample time to draw them in good style and we could study them together.

Please remember me to Col. Vaughan and Mr. Culbertson. I had the pleasure of meeting Mr. Culbertson last spring in Washington City. I hope you are all in good health and enjoying yourselves nicely.

Scientific men both in the east and west give Col. Vaughan much credit for the course he is pursuing in regard to the Natural History of the far West. At the same time he is attending to his duties as a government officer, in a very able manner, he will be doing much for the advancement of Science.

Dr. Shumard tells me he has seen a late number of the American Journal of Science in which Prof. Dewey has described several species of Carex from your collection on the Upper Missouri, one of which is Carex Haydenii.

I have formed a very high opinion of Prof. Swallow since becoming acquainted with him. From what I have seen of his manner of conducting the Survey, I am satisfied if the Legislature make the necessary appropriations, the work will be done in an able manner.

Yours etc.

F. B. Meek

P.S. Prof. Hall writes me if I had remained in N. Y. I would have had an opportunity to go with Maj. Emory to New Mexico on the new Boundary Survey. He said this position [?] is still open if I wish it [?]. Of course I can not leave now.

Prof. Swallow asked me the other day if he could get you in the spring when you come back. I told him I did not know, but presumed he could.[33]

Hayden's equally lengthy reply to Meek includes the following:

Council Bluffs, Oct. 6th, 1854.

Friend Meek,

...I received a very interesting letter from you at Fort Pierre. It says you are in the Geological Survey of Mo. and getting the snug little sum of $100 a month. Good! Just what you deserve and more if you can get it. I cannot tell you how glad I was to hear it. That will be a little better than $400 a year and get no credit even for your drawings...

As to your describing my collection with me, I have already spoken of that to you. I know of no one in whose judgment and accuracy in these matters I have greater confidence and I shall feel it a privilege if you will join me. I shall spend the winter at Fort Pierre writing out my notes, studying fossils and labeling them more carefully and packing them for their destination.

Let me ask you a few questions. Where had we better describe them? At St. Louis, or at Philadelphia? The transportation will cost nothing to St. Louis. I must have at least 50 different species of [fossil] plants. Can we describe them? Or must I send them to Dr. Newberry? If we can, I think we had better do it. Many of them are similar to those described by Hall in Frémont's Report. But many are of entirely different genera. Let me know what you think of these questions in your next...

You say Prof. Swallow asks you if he can have me on my return. Please present him my kind regard and tell [him] I would be pleased to do so... If the inducement is sufficient, I will leave the mountains after next summer. If not I think I shall go first to Bad Lands, then up to Blackfoot and as much into the mountains as I can go.

I saw a letter from Prof. Swallow on the Geological Survey in the newspapers which pleased me much. It showed Prof. Swallow to be a man of decided ability. I am well pleased with the course he is pursuing.

I will write to Shumard before I leave here, also to you again...

Ever yours,

F. V. Hayden[34]

Hayden did write Meek again, from Bellevue on November 11th, and in this letter he stated:

Now I wish to say to you that if Prof. Swallow will offer me a fair induce-

> ment I will come into the Missouri Survey sometime next season, perhaps early in the spring or summer. The Indians are so bad now, and are likely to grow worse, that I fear my operations must close, much to my regret. If Prof. Swallow offered me anything like a tolerable salary I would come...[35]

However, Hayden's operations were not to be terminated so soon. In the course of 1855 he would again find opportunities for far-flung travels by both land and river; and more than a year was to elapse before he would finally go back down the river to St. Louis. Nor would his eventual return to the States be for the purpose of coming into the Missouri Survey. Rather, it was to gird himself for a renewed program of western exploration, this time in the capacity he had long desired, as a naturalist "attached to government."

Thus the first main chapter of Hayden's work in which he was identified with Colonel A. J. Vaughan, was still far from completed when he wrote his letters to Baird and Meek in the fall of 1854.

12

Visit to the Settlements

In 1853, when Meek and Hayden first went up the Missouri River, it was the political boundary dividing the organized regions on the east (including Minnesota Territory and Iowa) from the unorganized domain or "Indian country" on the west. The latter region extended from the Canadian line southward a thousand miles to Texas. It was drained by the headwaters of the Missouri, and that river's other affluents from the Rocky Mountains and the Great Plains: the Yellowstone, the Little Missouri, the Cheyenne, the White, the Niobrara, the Platte, and the Kansas — to mention the larger of the streams by the names in present use.

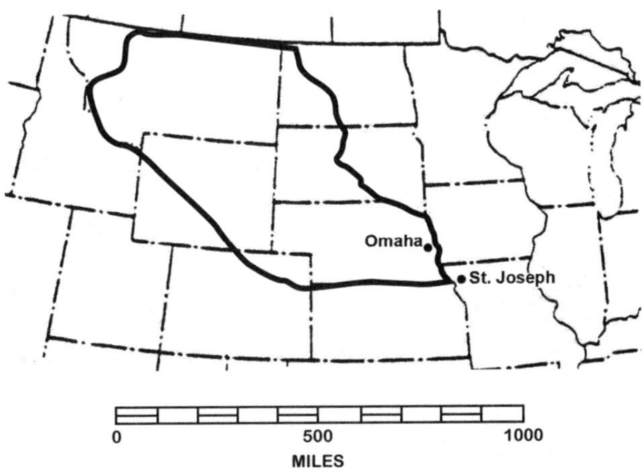

Nebraska Territory as of 1854, with later state boundaries

Before Hayden returned to the West a year later, this time in company with Vaughan, important events had taken place affecting the vast trans-Missouri domain. These events had culminated on May 30, 1854, the day before the *Genoa* cut loose from the pier at St. Louis for its spring trip up the Missouri. That was when President Franklin Pierce affixed his signature to the Kansas-Nebraska Act by which Kansas Territory and Nebraska Territory were created out of the unorganized region. Nebraska Territory was several times larger than the present state of Nebraska. It included all of the country from the 40th parallel northward to the 49th parallel, and from the Missouri River westward "to the summit of the Rocky Mountains."

As a consequence of these changes, when Hayden brought down the express from Fort Pierre in the fall of 1854 he found the little Nebraska settlements on the Missouri River astir with activity and excitement. As he wrote to Baird, "I am now in the midst of a tempest of politics, taking no part, however, myself."[1] In a subsequent letter (undated, but written not many days later) he wrote Baird, "Deliver me from Politics!" but actually he was already caught up in the swirl and, far from resisting it, swam along with the current vigorously and willingly. For, as he was quick to see, the political fomentation attending organization of the new Territories provided a golden opportunity to arouse interest in a geological survey, properly organized *under government.*

Though young (aged 25) and politically inexperienced, Hayden became a persuasive advocate of science in the service of government. True, his efforts did not meet with success immediately, but this was to be expected since he himself could not remain on the scene long. After urging his proposals as strongly as possible, he had to entrust them to others. Yet the experience initiated him into activity of a kind which would claim his best years and make him, in the words of J. Peter Lesley, "for a long time the master of the scientific situation at Washington."[2]

Hayden's bid for support of a geological survey did not claim all his time during his brief visit to the settlements. He seems to have crowded these precious days full to the limit with various activities. At Bellevue he started a little museum of natural history, doubtless the first institution of its kind in the Territory, a modest predecessor of the great museums which in after years would be founded in the region and similarly would emphasize its fossil treasures. His enthusiasm was contagious; others cooperated in the museum project and aided too in making collections locally. Starved for news after his months spent in the wilds, he devoured the papers and whatever other reading material he could lay his hands on (evidently not much, and scarcely anything relating to "the world of science"). Hungry for society too, he sought out old acquaintances and made many new ones. No longer counted a newcomer to the West, Hayden shared with others the experiences and observations of two journeys up the Missouri, good food for conversation, relished especially among those of more limited experience. And, fortunately, he took time to resume his correspondence. But for Hayden's letters to his friends "in the states," one would know little about those busy days in November 1854 before he set his face northward again to return to Fort Pierre.

November 11th, in particular, was "letter day" for Hayden. On that day he wrote not only to Baird, as already noted, but also to Meek, Engelmann, Leidy, and perhaps to others.

To quote further from Hayden's letter to Baird: "Nebraska is now being organized, legislature and council formed. Influential men tell me among the first things they will do is to petition for a Geological Survey of the territory. If they succeed, they say I shall have it. I would be glad to explore under such auspices. I think I could do things now to my satisfaction. I have commenced the nucleus of a little museum at Bellevue and got some young men interested in adding specimens to it. Perhaps that will do some good."[3]

To Meek, Hayden wrote:

Bellevue, Nebraska, Nov. 11th/54.

Friend Meek,

I wrote a few days ago in something of a hurry & now I am about to start for Fort Pierre. I thought I would drop you a few words more.

Coming down from Fort Pierre, a few days ago, though in a great hurry with the express I made some notes of the Geology which added to yours I think will be of some value. The river was extremely low and I found some beds that we did not see last summer. I also found several new species of fossils, a fine coral at Dixon's Bluff, etc. At mouth of White River I found some interesting bones, at lower end of Big Bend left bank I found a heap of chalk fossils, one or two new species I think.

I have been spending some days here collecting Carboniferous fossils. Got some fine ones. All of these I am going to claim as my own. I will send you my notes taken from Fort Pierre down when I return.

Politics run high here now. They are about sending up into the mountains to get the mountain vote. The territory is about to be organized and several influential men say that among the first things to be done will be to have a geological survey of the territory and [they] say that I shall have the survey. So much for promises. At any rate I have got up an interest in the matter. Something, I have no doubt, will be done of the kind…

As ever yours,

F. V. Hayden[4]

To George Engelmann, Hayden wrote a lengthy letter, the most interesting parts of which follow:

Bellevue, Nebraska, Nov. 11th, 1854.

Dear Dr. Engelmann,

I received a fine package of letters at Fort Pierre by the fall express, in which was enclosed a very kind letter from you. I was delighted to get some letters I can assure you, not having heard a word from any of my friends for 4 months…

I am glad also that you received a collection of plants from Prof. Hall. They are not as well preserved as I could have wished. You are aware that Botany was not the object of the trip and consequently my opportunities were very poor. The label in each paper applies to all in that paper, that is, they were collected or grew in or near the locality mentioned. Prof. Hall seemed to attach no value to the plants, and when I was dividing them into sets last winter he wished me to spend but little time on them as they were of no value, fossils being the only thing he cared about. So I got rid of them the shortest way possible. I hope, however, that you will be able to do something with them.

I have a fine [collection] at Fort Pierre, many different from those of last year and I think better preserved. I propose to take most characteristic specimens of each species, number them 1, 2, 3, etc., give place, date, etc. and the geographical distribution in the West. Thus No. 1 on label refers to notes with full remarks to soil formation, etc. I feel anxious to determine the geographical distribution of plants in the Upper Mo. Country. When thus prepared I wish to send them to you in spring by Col. Vaughan so that you may describe any new species.

I send you a box containing three bottles of alcoholic specimens which I wish to have sent to Prof. Baird. In the box you will find all the Cacti I have collected this season with notes, etc. I think there is one new species, found only on the Yellow Stone, called Jumping Prickly Pear, because the seed vessels fly on the traveler like burrs. Description in field — "Nearly circular leaves, flat, branching, spines profuse, two kinds, one or two large spines sometimes 2 or 3 inches in length, reddish brown, surrounded by 6 or 8 small white spines. Seeds always white... Very barren soil, quite abundantly in patches. Seed vessels numerous on a plant, — three to six." Again, "I noticed on upland basin a very large bunch of Jumping Cactus in which some of the leaves were six inches in length, four [inches] broad, three-quarters [inch] thick. On several leaves I counted eight seed vessels. Spines around edge very thick and red." This was a rank growth. These notes are from my note book.

I think if you get the box safely you will find good specimens of O. fragilis. There are two seed vessels corresponding to your figure. You will also find some small specimens of petrified wood. Keep them if you like them. Also some pine cones. All from the Yellow Stone. The box will be sent to St. Louis care of P. Chouteau, Jr. & Co., and will come to the office within a few days after the arrival of this letter I think...

Perhaps you wonder at receiving a letter having the above date and place. On arriving at Fort Pierre there was no one to whom Col. V. wished to entrust his express. The excited state of the Indians rendered it unsafe for me to operate much during the fall. So he wished me to go down to Council Bluffs and see all his papers safe in the Post Office. I have just arrived, 15 days going, in a small skiff, low water and much detained by wind, etc. I shall go back soon on horseback by land. A cold time I am afraid. But I have all I can do this winter in stuffing animals, preparing my notes, spec's, etc.

I made a good collection on the Yellow Stone, about three tons weight. Many new species, especially fossil plants. I am inclined to think that the Ancient Flora rivals the Modern or Recent. Beautiful!! I have quite a variety of stuffed animals, as — 4 beavers, 8 prairie dogs, 2 grizzly bears, 2 buffalo calves, 3 antelopes, a large number of heads of antelope, deer, big horn, etc., with skins and heads [and] about 10 inches of neck, making fine specimens. 2 skunks, foxes & wolves and a variety of birds, etc. A good many alcoholic specimens also. So you see I have already a pretty good collection for one to make single handed & alone.

I am very much afraid recent trouble with the Indians is going to prevent me from going to Bad Lands next spring and perhaps suspend all my operations. Much trouble is expected.

Please write me as often as you can and give me the news in Science. I discover I can hear nothing. It makes me uneasy.

Is government doing any thing? Are any expeditions being fitted out for exploration? Etc. Please remember me very kindly to Mrs. Engelmann, and accept the

Kindest regard of

F. V. Hayden[5]

To Leidy, Hayden wrote:

Having just arrived at this point from Fort Pierre I thought a few words relative to 'bones' may interest you [and he devoted most of his letter to this subject, concluding by saying] it has been very dangerous working, for me this summer, increasingly so, but I have got off safe with one or two little skirmishes with Indians. Government must do something with these Indians or all traveling here is at an end.[6]

Because of his interest in the political situation, and his own involvement in it, Hayden stayed on at Bellevue for a couple of weeks, and during that time worked energetically for the proposed survey. The following is from a letter to Baird written at Bellevue:

Bellevue, Nebraska, Nov. [17, 1854]

Prof. Baird,

My Dear Sir,

I have already written you a long letter but some things have turned up since which I wish to lay before you and obtain advice.

Coming down from Fort Pierre, I had the good fortune to find some good petrifactions, Cretaceous fossils and vertebrae, etc. I have shown them to many, who seem to be much interested. I have also started a nucleus for a museum for the territory, to which I contributed some of my [vertebrate] fossils and a great variety of Carboniferous fossils which I have gathered since I have been here. I have labeled and arranged [them] in the house in which Gov. died & which, if the capital is located at Bellevue, will be an important point. An intelligent young man of property, a brother of the most prominent candidate for Congress, I have interested to take care of it and add to it.

All the citizens of the territory without respect of party are very desirous of having an early survey of the territory, and when the Legislature & Council meet, which will be this winter, one of the first things they propose to do is to memorialize Congress to make an appropriation for a Survey. The delegate will also urge the matter strongly at Washington. The different papers are also going to urge the necessity of a Geological Survey, etc. I shall write several articles setting forth the importance of such a movement and establishing a territorial museum. The most influential men in the territory say that I shall

> have the charge of this survey. Now I would ask you what would you think of the prospect under such circumstances? Do you not think if all the people wish it and the delegate asks it that Congress will make an appropriation? Perhaps by some political maneuvers I would not be selected but that is not the question with me. Could such a thing be done? ...I shall labor pretty hard for such a thing to succeed even if I have no position in it, for I am convinced that Science would be very greatly benefited.
>
> The three bottles of alcoholic specimens will start by stage in a few days... I think you will find some things among them that will please you. I have the copper can filled full which you will receive in the Spring. These will be labeled separately. The bottles I found impossible to label properly for reasons explained in a former letter. Please write me all about them. I have found a fine lot of Cacti for Dr. Engelmann. One new species I think.
>
> I have now been here nearly two weeks awaiting political movements. There is a question whether the mountain vote should be excluded or received...
>
> If I had time I would make some money this fall hunting for coal. Coal is the great desideratum of the territory, and people are confident they have it in abundance but I fear not. A little may be found but no very valuable beds...
>
> Yours most sincerely,
>
> F. V. Hayden[7]

By late November, time had run out for Hayden, and his final letters from Bellevue were written on November 26th. To Engelmann he wrote about the arrangements he had made for getting his collections down to St. Louis.[8] To Baird he again stressed his conviction that "the people of Nebraska are very anxious to have an early Geological Survey of their territory," stating that the citizens found its advantages "so obvious that everyone entered into it immediately, and this winter the Legislature and Council will take positive action in respect to it. Three newspapers here will advocate it in the right time... I have even the positive assurance that next Spring the appropriation will be made. However, if it is made in one year from that time I shall be satisfied..."

In a later paragraph, Hayden wrote, "At the suggestion of Col. A. R. Gilmore,[9] I have prepared a small box of nice specimens well labeled for Senator Douglass[10] [*sic*] which he will carry to him and press the matter of a Survey. Col. Gilmore is the most influential man in Nebraska, an intimate friend of Senator Douglass [*sic*], and a train of influence may be reached then which will effect much, this administration. Col. Gilmore is confident that there will be no trouble in drawing an appropriation. He will call and see you as soon as he comes to Washington... He will explain to you the nature of the enterprise, and I trust the plan will meet with your cordial approbation..."

Hayden also wrote Baird in this letter about the arrangements he had made at Bellevue "to secure the fishes and reptiles of this region, by leaving bottles of alcohol in various places," so that some of his interested friends might have the means of preserving specimens. And in a postscript he added, "If you think all is right I feel very confident that your influence and exertion at Washington would bring about the desired result, and open the way for the exploration of one of the most interesting conditions in the world. I need not tell you how much I am interested in the success of the enterprise. I pray you to examine the matter, and if the prospect is as fair as my imagination paints it, to have it done. That you must feel a much deeper interest in it than I, I can have no doubt, for you can better feel its importance… So I am now, my hands are tied as it were, having to go in the midst of dangers unarmed and alone, with no one to assist me."[11]

Hayden's letters leave no doubt as to the great "lift" in morale he had received as a result of his visit to the settlements and from the encouraging outlook for a future geological survey of the Nebraska territory. In closing his letter of November 26 to Baird he wrote, "I shall now face the cold north winds on my way to Fort Pierre with pleasure and a glad heart. I have made all my arrangements here to start tomorrow across the Big Sioux and in 15 days I hope to be at work in my collection."[12]

In the course of his northward journey, Hayden wrote Baird once more on the subject of the survey, which he declared was uppermost in his mind. The letter, of which only the first page is preserved, was written from Sergeant's Bluff, and dated Dec. 6, 1854, and Jan. 14, 1855 (the latter probably the date of mailing). In this letter Hayden reported to Baird, "On my way to this place I stopped two days at Kanesville and laid my ideas before Dr. Louis, a friend of mine who is government land receiver. He is a man of much influence and assured me that there would be no trouble in securing an appropriation. He has already written to Senators Jones[13] and Dodge[14] of Iowa, also Hon. Bernhard, senior member of Congress from the Kanesville District. These men have represented the territories for many years, and Dr. Louis thinks that the influence of these men now would secure the desired object. Mr. Chapman, formerly of Ohio, will undoubtedly be the delegate to Congress from Nebraska. He has promised to make the matter a point. All fully appreciate the importance of such a movement in advancing the settlement of the territory, and the great want of fuel leads the people to inquire earnestly whether Nebraska contains coal."[15]

Having finally left the settlements behind him, Hayden could reflect at more leisure on this, his first experience with "the tempest of politics." That he came to take a realistic and sober view of prospects for an early approval of a survey is clear from the one additional reference which he made to the subject in the letters written that winter. Writing to Baird from Fort Pierre on February 9, 1855, he states, "I wrote you last at Sargent's Hill, and perhaps I have written oftener than was necessary, even so as to annoy you, but I hope not. While below, the people seemed so much interested in a Survey that I thought such a thing possible, but promises in respect to Natural History are more easily made than fulfilled. However, a survey may be got up, within a few years. Perhaps when Dr. Evans returns from Oregon he will devote his attention to it."[16]

As it proved, Hayden's efforts on behalf of a geological survey were not just wasted energy. For one thing, they served to give shape to his ideas concerning such surveys. Furthermore, he had succeeded in implanting these ideas in the minds of other men, where they took lasting hold. As he persistently worked at the proposal, year after year at every opportunity, it matured and eventually bore fruit. Thirteen years later (in 1867, two years after the Civil War, which had interrupted almost all activity in the West) Hayden's dream became an actuality, when the General Land Office of the Department of the Interior authorized a Geological Survey of Nebraska, with Hayden as geologist in charge. The Survey of Nebraska marked the beginning of the "Era of Territorial Survey," which continued until 1879 and was a highly significant chapter in the annals of American science and the history of the West. In a very real sense this era can be said to have had its inception as early as November, 1854, when Hayden, during those busy days at Bellevue and the neighboring settlements, first sought zealously to win authorization for a geological survey of the newly created Nebraska Territory.

13

First Winter in the West

No further details are known about Hayden's journey "home," but evidently his glad heart, and his anonymous steed, sustained him on the return trip to Fort Pierre, which he reached about the middle of December. There he kept busy during the remainder of 1854, through Christmas and New Year's, festive occasions which, presumably, did not go wholly unrecognized even at the forts on the Missouri.

In the long months of that winter, Hayden's heart was not always glad. For one thing, it is evident from his letters that he soon felt the isolation of life at Fort Pierre, probably much more so than did most of the men who wintered there. In writing to his friends he pleaded for news from the outside world if and when letters could be gotten through to him.

Thus on New Year's Eve 1854 he wrote Engelmann, "I feel very anxious to hear something. I wish you would take the trouble to inquire how they [letters] may be got to me most expeditiously." Also, "The members [of the Fur Company] at Fort Pierre were sorry that I did not go to St. Louis. I am sorry now that I did not go, but they gave me no orders to that effect… I expect that there will be no express sent up this winter. Perhaps Mr. Campbell may be sending up… When I shall return I know not, perhaps in the summer with the boat and perhaps not until fall."[1]

Yet, as always, he was full of plans. He reiterated his hope that Engelmann would be willing to join him in preparing "a catalog of all the plants of Upper Mo. with their geographical distribution. I should feel it a great privilege if I could join you in making out such a catalogue. It might be a contribution to Botanical Geography." He assured Engelmann that in the spring he should be able to collect several hundred more plant specimens for him. These he would send down to St. Louis by Vaughan, "with such notes as I shall be able to make." He promised that "I shall watch the Cacti more attentively this spring and doubt not that I shall catch them all in bloom."[2]

Earlier, Hayden had written Baird of his intention to work on his notes and specimens during the winter he would spend at Fort Pierre. That he had other purposes in mind as well is first intimated in the letter to Engelmann. He wrote, "In a few days I go out to the Cheyenne, also to the 'Bad Lands' and get a turtle for Col. V. to take down with him in the spring. The Indians are not well disposed at present and there is considerable danger apprehended, especially from the Brules who were in the Platte affair. I have, however, concluded to spend two or three months next spring in the 'Bad Lands' with a couple of men."[3]

One familiar with the Dakotas might at first view Hayden's proposed plans for winter trips as dubious if not downright foolhardy, for even under the best of conditions it would be difficult at this season to make geological observations and to collect specimens from the frozen ground. More serious would be the risk of encountering blizzards. The region is notorious for vast storms that sweep in from the Rockies and bring blinding drifting snow. In their wake comes paralyzing cold; temperatures may plunge to 20 or 30 degrees below zero, or lower. Such weather can halt outdoor activity abruptly and sometimes disastrously. Even now, field work is generally impracticable during the winter months, though the present-day scientist has advantages undreamed of in Hayden's time, such as weather predictions that give forewarning of storms; swift and comfortable transportation; and the necessities of life (including medical help in case of emergency).

Yet it was not through ignorance of winter weather nor reckless disregard of it that Hayden ventured afield during the early months of 1855, both in the immediate vicinity of Fort Pierre and to destinations far distant. Indeed, he left the shelter of this post *because of* the winter's severity, since it meant Indians were less likely to be abroad, particularly in exposed places like the Badlands. Cold, snow, and hunger might be encountered, to be sure, but hardships seemed a price worth paying for freedom to go where he wished, and an added measure of safety from the marauding Brulees. These were factors not to be ignored after the recent Indian troubles. Hayden was conspicuously fearless even by the standards of the frontier, and it was often remarked that he seemed to enjoy a special, almost magical immunity from molestation by Indians; however, he generally took what precautions he could.

There were other considerations as well. Some of his trips were evidently timed to coincide with those of the bourgeois of Fort Pierre, Charles E. Galpin, whose assignments, frequently difficult and hazardous, took him away from his post even in the dead of winter. In Galpin, Hayden found a hardy, experienced companion and a kindred spirit. In addition, Hayden probably reasoned that while based at Fort Pierre he should accomplish as much field work as possible in the areas within reach of this post, especially the Badlands, so that by summer he would be free to seize any available opportunities for longer journeys, such as a return to the Upper Missouri region.

Hence the strenuous activities of Hayden during the winter and spring of 1855 However, after taking these factors into account, one may still marvel that Hayden could induce his helpers to go with him, that his parties survived their experiences as well as they did, and that Hayden's early failures to accomplish what he wished did not deter him from repeating his attempts. There are no known records of his trips other than the terse accounts in his letters. In reading these sketches, one's imagination must supply the details. Only so is it possible to appreciate the hardships and dangers involved in Hayden's treks to such places as the Badlands and the Black Hills with little food and oft-times no shelter; or the difficulty of getting collections hauled or packed back to Fort Pierre. Considering what these collections cost him in labor and risk, it is small wonder that, in his letters, Hayden often expresses anxiety concerning the safety of the glass containers, frail bundles, and makeshift wooden boxes and barrels in which his specimens had to be entrusted for shipment to St. Louis and Washington.

Writing to Baird from Fort Pierre on February 9th, Hayden briefly mentions a trip already made, probably the first of his longer expeditions out of the fort, and he alludes to another to begin on February 11th. He had been relatively fortunate on this first trip as regards weather and therefore quite successful:

> 9 Feb. [18]55
>
> I have just returned from the Moreau Trading Post, 14 days travel from Fort Pierre. I had Mr. Galpin as companion, who feels much interested in geology and has a pretty good practical knowledge of it. I had a fine opportunity for examining the Fox Hills, Moreau River, and many other places of much interest, and made a fine collection of shells and large bones. In two days from this time I start for the "Bad Lands" to spend a month, with one cart, four horses, and two men. There is not as much danger now, and the Brules have threatened to kill any white man who may fall in their way. They are now supposed to be on the headwaters of the L'Eau qui Court. It is a hard season... we have as yet had but little snow, but some very cold weather.[4]

In this letter and others, Hayden refers to disagreements between Vaughan and himself which were threatening to disrupt their partnership. He also sets forth the terms of an offer of financial assistance which Charles E. Galpin had made him. Galpin's initial interest in matters of science, it appears, had been stimulated through his contacts with Hayden at the fort and on their trip to the Moreau Trading Post. Galpin was evidently a man who could enter into Hayden's enthusiasm for scientific discoveries for their own sake; at the same time, being a trader, he could share Vaughan's view that if scientific collections were saleable they might well be made a source of profit. Undoubtedly Hayden was predisposed to favor an alliance with his new friend. Galpin's kindliness would have been a welcome exchange for Vaughan's intransigence; and an official connection with a leading member of the Fur Company which operated throughout the Upper Missouri region would offer distinct advantages. At the time, Hayden still had no assurance of government support or of any other backing which would help solve his old problem, how to sustain himself and yet remain at work in the West. After a year of roughing it he had become, as he put it, "rather destitute," a description which was probably an understatement considering his poverty to begin with and his dependence on Vaughan for virtually everything. Although nothing came of "the Galpin plan," its pros and cons as stated in Hayden's letters provide an interesting example of the relationships which existed during this period between scientist-explorers and such frontiersmen as fur traders, guides, and Indian agents. As Hayden said in his same February 9, 1855 letter to Baird:

> I have just had a more just and amicable arrangement with Col. Vaughan. His idea of right and wrong took altogether too wide a range in some respects and in others too narrow. So I requested a written agreement. As it stands, all peltries hereafter obtained with some already obtained, all plants, all alcoholic specimens & birds we mutually present to you. The petrefactions are to be taken to St. Louis and there arranged and described by myself or at my request. This I wished to have to prevent the undescribed species from falling into the hands of any one & no credit reverting to the laborer.
>
> A proposition has been made to me which I wish to lay before you, and desire your opinion as to its feasibility. Mr. Galpin, the bourgeois of Fort Pierre, has made me this offer and is desirous of carrying it through, and if there is some prospect of government making an appropriation it would be well to consider it perhaps. He will furnish $1500, or more if necessary, to operate 18 months, and [this] with other facilities which the Fur Company can afford would be amply sufficient. He puts that capital against my time. He is willing to wait for a return of his money 18 months after the collection is finished if it is necessary, but wishes to realize a fair percentage on the capital invested. I told him I would write and get your opinion, and if the plan looked favorable to you I would enter into it heartily. Could two or three thousand dollars be raised by subscription among scientific men & colleges? Would government be willing to purchase such a collection? Or could it be divided into sets and sold in Europe for anything worth while? The whole Upper Mo. company would become more or less interested in procuring specimens and Mr. Galpin thinks that stuffed animals might be obtained to the amount of $1000. Their teams and men are ranging all over the country east of the mountains and all expenses would be comparatively light.
>
> I believe I have given you the substance of the matter. Please give me your opinion as soon as convenient. If there is a fair prospect of a survey in a year or two, it would be better, perhaps, to let the opportunity pass. Of this you will know best.
>
> I think I shall return in the steamboat next Summer, and then I hope there will be some opening by which I can labor and receive a salary, for I am getting rather destitute, and will be far more so. Mr. Galpin with genuine liberality has offered to let me have any amount of money I may wish, to appear respectably on my return, for which I feel toward him the deepest gratitude...
>
> My notes have accumulated to a large quantity. I have just copied my Geological Sections, and they fill over a hundred pages of letter paper, closely written. I intend to explore to some extent [the] White, Cheyenne, Moreau, Grand Rivers, etc. and secure a set of fishes from each for you. The birds are very few here, especially in winter. I hope in Spring to get a good many...

I have not heard from you since your letter of June 3rd. I feel very sorry for I stand in need of your counsel very much. I think there must be an express up before the steamboat comes, though it is uncertain.

I ever remain with sincere regards,

Yours, etc.

F. V. Hayden

Since finishing my sheet, we have heard the sad news of the murder of Mr. Malcolm Clark on the Yellow Stone by a war party of Sioux.[5] You are doubtless familiar with his name. He was a fine man, a gentleman in all respects, once a cadet at West Point. Dr. Evans and Gov. Stevens will recollect him well. All the Indians are very bad and daily growing worse. It is not impossible that I may not live to return to the States. In case I should not, I wish you would take my collections and dispose of them as you may see fit. My notes and journals I have left in the care of Mr. Hodgkiss, which I wish you to receive also. The Indians are much opposed to anyone's getting specimens in their country, and will yet make trouble with someone.

I will write you again as soon as I return from the "Bad Lands," which will be in 4 or 5 weeks, when I expect Mr. Culbertson will pass here on his way down.

I shall have a chance to send you something in the Spring, I hope.

F. V. H.[6]

Writing to Joseph Leidy on the same day, Hayden restated Galpin's proposal, asking for Leidy's views as to its practicability and venturing the opinion that "Such an opportunity to explore the country thoroughly cannot occur often, unless a survey be made." The other parts of the letter to Leidy follow:

Fort Pierre, Feb. 9th, 1855.

My dear Sir,

Perhaps you may be interested to hear how I am succeeding in my explorations, especially in accumulating bones.

Before leaving St. Louis I received a very kind letter from you, and I should be glad could I hear from you again. Any instructions that may occur to you that may facilitate my collection of specimens I would be glad to receive from you.

However, I may not be able to hear from you until the boat comes up, and it is my present intention to return in the boat. The danger is too great and is constantly increasing. We have just heard of the murder of Malcolm Clark, one of the partners in the Fur Company, killed by a war party of Sioux in ascending the Yellow Stone. Several murders have already occurred and some bands have threatened to kill any white man they meet.

In two days from this time I start for the "Bad Lands" with a cart, 4 horses and 2 men. There is no snow on the ground now and may be none this month, and no danger is apprehended from Indians, whereas in the months of May and June it would be almost certain death for a small party, and much trouble for a large one.

As yet I have obtained but few bones, but what I have will I think be of much interest to you. Two sets of the bones of a Saurian animal from Cretaceous formations, which may be the same as the one spoken of by Nicollet, but much more perfect. Some few specimens of fish, one though small I think will prove a most interesting species as having a close resemblance to existing species...

Mr. G[alpin] has a good practical knowledge of Geology and is a man of no ordinary intelligence, and withall one of the kindest hearted men I ever knew. Of this fact Dr. Evans can testify...

I trust I shall see you during the coming summer.

Ever yours with sincere regards,

F. V. Hayden[7]

The next day, February 10th, Hayden wrote to Meek. Parts of this long letter are worth quoting:

Fort Pierre, Feb. 10th, 1855.

Friend Meek,

...I have just returned from a trip to the Moreau Trading Post, where I had a fine time, made fine collections, and determined some interesting points in Geology. I obtained many new spec's especially in Fox Hills, also some fine Cretaceous plants, fish & scales, and Coleopterous insects. Also some very large leaves, together with their geological position...

I soon start for "Bad Lands" where I hope to make some fine collections, there and on the Cheyenne.

Mr. Galpin has acted the part of a friend toward me and is continuing to do so.

I shall get many fine specimens from the different trading posts.

I hope you will be in St. Louis at the time the steamboat arrives. It is my present intention to return then, and I shall wish to examine my specimens there. I wish you would have an eye to it and manage to have a few weeks of leisure time in July or August. We will then have a room and look over such specimens as are new and suitable to figure, etc.

I had a tearing up of matters with Col. V. He began to talk wild and wanted to have the game in his own hands, so I called for a written contract, which we have.[8] He incurs all expenses to St. Louis. I then have the privilege of arranging and describing. All peltries, plants, birds, alcoholic specimens, etc. go as a mutual present to Prof. Baird. This I think is right.

I wish you would be securing books by purchase or loan that will help us. You may depend on looking at a fine collection and a plenty of specimens for comparison...

I think the Cretaceous formation extends nearly to Council Bluffs. I found two undoubted Cretaceous fossils at Sargent's Hill this fall, etc.

I sincerely hope you will have leisure to attend to them [the fossils], and I think we ought to get up an important paper. I have just been copying my sections. I have already over 100 closely written pages of letter paper. These will interest you. I have still several months to labor and in that time I hope to get at least as much more. I have got and shall continue to get a fine set of geological specimens, soils, etc. I have also a good collection of recent shells. I hope to give Sage Creek another search and perhaps something new may be found as we spent only a few hours there. That point near Pinoe's Spring is a place of much interest to me.

What do you think my collection will bring me? Anything worth while? Please inform me something about it when you receive this letter. I may not be able to hear from you until the boat comes up, but let me know even then.

I hope the Survey of Mo. will continue. I hear that Prof. Swallow is pressing it forward with great energy and I have no doubt that if he is well sustained by the Government he will bring about most important results...

We have just heard of the murder of Mr. Malcolm Clark, a partner in the company. He was killed by a war party of Sioux while ascending the Yellow Stone. They are constantly knocking someone on the head...

I will show you all my notes and journals on my return, and those with my specimens will enable us to determine the Geology east of the mountains...

Tell me all the news. What surveys are started, and what is going on in the scientific world? Is Ohio doing anything? What position is it that Shumard is expecting? Give him my regard. Has Dr. Evans returned from Oregon, and when will his report appear? Remember me kindly to him if you see or hear from him. I shall rejoice to meet you in St. Louis in July. Perhaps I may lose my top knot before that time. I can assure you it is scary times here. How is your health now? I hope it is improving. I never had better health, eat everything I can get.

I remain ever yours sincerely,

F. V. Hayden

I have not heard from you since September.[9]

A month and a half elapsed before Hayden was once more back at Fort Pierre and could resume his correspondence. On March 25th he wrote to Meek and in this letter reported a long journey which had become a grim struggle for survival against conditions far more adverse than those encountered on the earlier trip to Moreau River:

Fort Pierre, March 25th, 1855.

Friend Meek,

I feel anxious to let you know my whereabouts and my operations, though I have not heard from you since September. Since that time I have written several letters to you which I hope you have received. Are you now with Prof. Hall? If you are, your chances for the study of Paleontology must be very great. Please look over the Cretaceous and Tertiary formations very carefully and collect as many works as you can. I hope you still adhere to your wish to describe my collection with me. I depend on it and feel very anxious that it should be so. I have very much confidence in the accuracy of your opinions in matters of Geology and fossils. You may rest assured that no collection ever brought from the west side of the Mississippi equals it in any respect, or rather will not when I get it safely to St. Louis. It is very full. We can have as many as we please for comparison. There [are] I think at least one hundred new species, and finer specimens of many that have been described than any I have ever seen below.

It is my present intention to return with the boat, which will get to St. Louis, as you know, about the 1st of August or last of July. Then I wish you could get a short furlough from Prof. Swallow, and we can examine it [the collection] and select those [specimens] that will be suitable to figure or make such arrangements with it as you may judge for the best. It will probably be stored in a vacant room in the Superintendant's Offices. Col. V. says there are several rooms vacant. At any rate it will be for the Col. V. to secure a good room. I have a written contract with him, by which I have the privilege of arranging and describing them. He seemed to think that would detract from their money value, so I was for backing out of the whole thing and so he agreed to that. He was inclined to vary somewhat from his verbal contract and keep too many choice specimens for his own use.

Mr. Galpin has been very kind to me, in fact like a brother. You are mistaken [in] him. You thought him rather hard, but you know when they give they give, and when they charge they charge. The Lower House represented to Galpin that Hall was good for so much and they thought they would have it. But let a young man come up here poor they will help him. I owe Galpin very much. He is coming down this Spring with Culbertson, and Hodgkiss goes down in the steamboat. Galpin has facilitated my researches very much. He has himself a very good practical knowledge of Geology and is always with his hammer in hand. I have traveled hundreds of miles with him this winter.

I have just made an abortive trip to the Mauvaises Terres. The first of the winter was very mild. It continued so through the month of Jan. We all thought it the safest time on account of the Brules, so I started the 11th Feb. with two men and four horses and a cart. A storm overtook us at Grindstone Hill and then another at Bull Creek. Snow came 2 feet deep and the thermometer was 20° below zero. I got to Sage Creek, stayed there 5 or 6 days, and then started back. This trip proved a failure, though I got that big Ammonite that we hid in the brushes, and a splendid collection from Pinoe's Spring. I made a plenty of developments there which I am confident have not been made before.

I shall start again now in about 10 days and stay until I get through if it takes me three months. I intend to go into the Black Hills and the other side of White River, nearly to Fort Laramie, and then take White River to its mouth.

I felt much disappointed at being foiled by the snow, so I sent the men and team back, took one horse and pushed across the country alone from Pinoe's Spring to the [trading] houses on the Cheyenne. I got lost and was out three days in the cold & snow without anything to eat. This was rather hard. The third day at night I got to the houses and made a small collection. About three days time and then Galpin came along from the Fort, going out to the camp. We then went to the Baggage about 100 miles farther, just at the foot of the Black Hills, and there another snow storm overtook us. I made a small collection there. Visited the Black Hills, determined their geological formation & came back by way of Cherry Creek. They had nothing to eat then at the Baggage, and for 10 days I had not more than half a meal per day. We started away, Mr. Galpin, Mr. Zaphrin the interpreter, and myself, with only food enough for a couple of meals. We depended on our guns and could hardly kill a prairie dog. For four days we traveled 150 miles with nothing to eat but one prairie dog divided among three. We got to the Fort 22d March, as nearly starved to death as I wish to be. We all ate too imprudently and got terribly sick. I am just getting able to write. I was really afraid I would die any day for a time. Galpin is quite sick yet from the effects. I told you a man in this country must see hard times...

I find, in reckoning up, the distances I have traveled since leaving St. Louis 1st of June last to be nearly six thousand miles.

I think I shall go up to Fort Union with the steamboat again and verify some of my old sections. I have collected lots of facts in relation to the geology of this region which I am very anxious to examine and discuss with you. Do not show my letters to anyone and do not talk about any information I may convey to you in them. You know people are not good at giving authorities nowadays. It will be time enough when we have a chance to publish our investigations to the world.

You will see in the Indian Reports Vaughan's report. I wrote out that report and like a cussed fool gave him my discoveries on the Yellow Stone. It is doubtless published ere this.

Don't fail to write me when you receive this. I may not get a letter until the steamboat comes up, but you know it will be very acceptable even then. Write all the news. Are there any surveys in progress? If so let me know where and who has charge.

Remember I have many new species of Tertiary plants. I think we can describe them. If not, Dr. Newberry must, but I prefer that we should do it. They are all new species, I think, and I can refer them mostly to their genera.

I remain ever your sincere friend,

F. V. Hayden[10]

For a couple of weeks, Hayden pursued his work at Fort Pierre. His need to recuperate fully from the last trip probably had something to do with this. He and Vaughan were again on good terms and had agreed to continue their association through another summer. Hayden's decision to return to the Yellowstone with Vaughan in the summer is stated in his next letter to Meek.

[Fort Pierre], April 5th, 1855.

Friend Meek,

Since writing my letter to you I have altered my mind somewhat. In talking with Col. V. he is very anxious to have me go up the Yellow Stone with him again next summer. Now in order to complete my researches it is very important, for from want of experience I now see many things which I neglected. I shall probably return in September. My collection will go down in the steamboat and be stored away there until my return. Now what do you think of it? I can doubtless get many new species, correct and revise all my sections, and make any alterations I may see necessary. You know that the Yellow Stone country is terra incognita, and I believe I have the honor of being the first and only [geologist] who has made a minute examination of it. Still I believe I can get many new things yet. Be sure to send me by boat all the articles that have been published on this country since I left…

Monday coming I start for Bad Lands to stay six weeks. I have only one man and a boy, one cart and three horses…

Can you not obtain next winter to your own use? As soon as I come I would like to go to work and describe my collection and have it ready to publish in the spring. I long very much to see you, and if I do not get scalped I will before the close of 1855.

I am quite certain I today found two new species of fossil mammals. During this trip perhaps I will find something more. My Cretaceous fossils I think are the finest that ever met the eye of man. They are what you may call perfect.

Don't fail to get Cretaceous and Tertiary books as far as you are able, and try to be in St. Louis on my return so that we can just open the whole collection and select such as we wish to figure. I have a written contract with Col. V. in which I have it stipulated that I have the privilege of describing the whole. So I have the control of that. So you need not fear but that you will have material enough for comparison. On returning from the Bad Lands I shall slip into a Mackinaw boat and go to Bijoux Hill and examine that and come back in steamboat. So you see that will be nice.

I have sent my plants to Dr. Engelmann.

Yours sincerely,

F. V. Hayden[11]

Another letter written by Hayden on the same day is preserved. It was addressed to George Engelmann:

Fort Pierre, April 5th, 1855.

Dear Dr. Engelmann,

I have not been able to hear from you since last September, much to my regret. Since that time I have written you several letters and sent some Cacti, etc. to you...

I have just returned from a trip to the Black Hills in which I made some fine notes. On the south side of a lofty butte about 1200 feet high, a part of Black Hills, I found two species of plants just coming into bloom. One of them is a red looking plant, flowers on a spike. The other resembles a Hepatica and is called by the Indians the Navel Flower, from its resemblance to a navel, and when this expands they think that the unbilical [*sic*] cord of the young child Spring is severed.

I made an abortive trip to Mauvaises Terres. I started Feb. 11th but encountered a severe snow storm and was obliged to turn back after reaching Sage Creek within 12 miles of place of operations. I shall start again the 9th April and then stay until I accomplish my object. I have made something of a collection of fossils but other things are not much. This is not a good Fort [at which] to obtain skeletons. There [are] so many hungry mouths about here that an animal is broken up before it gets cold.

I would like to have you see Col. V. when he comes to St. Louis. You will find him at the Planters House, also Mr. Galpin and Culbertson who are going down.

We may hear from below before the steamboat comes up, but it is uncertain. I have intended to return with the steamboat, but Col. V. wishes me to take the tour up the Yellow Stone again, and I am inclined to do so. I shall be able then to get to St. Louis in September, and I wish very much to review my labors of past summers. In case I do stay up here I wish to ask some great favors of you. My collection will return in the boat, many boxes. A room will be provided by Col. V. but I wish you to see them all handled and put in the room, as I fear they will be misused. The boxes will not bear rough handling and I fear for them.

I shall also wish for some things. 1st, two kegs of alcohol, one of them medicated and the other not, both marked "poison." One ream of paper, same as I got. I wish one keg to take up the Yellow Stone and the other for a trip contemplated to the Black Hills but not yet certain. By an arrangement with Col. V. all specimens, except fossils, go as a mutual present to Prof. Baird and therefore I think that he will see that these things are paid for. Please send them on board the steamboat care of Col. V. If I do not need them I think I can inform Prof. Baird in time.

I feel quite anxious to go up the Yellow Stone again for several reasons, and as I will be gone only about two months I think it will pay well.

I send you a large portion of my plants which I wish you would examine... You will find some notes under the top bundles. I could not label them as well as I could wish, for want of paper. So I did the best I could with them, and hope you will find some things interesting. Some of them are common in the States, I know, but I suppose you wish to ascertain the complete flora of this region and the geographical distribution of all. I have endeavored to

> make the set as complete as possible and when you find a specimen not labeled or numbered it is because it has been recorded in some of the preceding papers. I did not like the plan of putting the specimens on one another, but want of paper compelled me. Please preserve them until I see you, when I hope to furnish you with many more…
>
> I am obliged to make use of very small means, and often, do as I can, make my pursuits secondary even. In a Government Survey or Expedition, where I could have the proper facilities, I could do things much nicer.
>
> Mr. Galpin, the present bourgeois of Fort Pierre, has been very kind to me, even like a brother to me. To him I am much indebted for whatever I have or may accomplish about Fort Pierre.
>
> I have directed a letter to Mr. Meek in your care. Please forward to him. I do not know where he is, but suppose he is with Hall.
>
> Col. V. just tells me that he will engage a young man to look after the boxes and see them stored away, but still I wish that you or Mr. Meek would have an eye to them, if you are not too much hurried. I am sorry to annoy you as much as I do, but I know your willingness to aid in all such enterprises…
>
> Do not hesitate to make any suggestions or give me any advice you may think I need, for you know my main object is the advancement of Science, and if I can do a little I shall be glad. Therefore I shall not only not be offended but feel that you are my friend indeed if you advise me in any way you may think best…
>
> I should like very much to get any works that may have been published on this country during my absence. Perhaps Prof. Baird will send them to me. It seems as if I was excluded from the world, an exile as it were, and I sometimes feel lonely.
>
> I send a keg and a bottle of alcoholic specimens & a small keg containing embryos for Prof. Baird. Should they come to you, I would like to have you put them into Adams Express and send them to Prof. B. Please accept my sincere gratitude for all your kindness, and believe me,
>
> Yours most sincerely,
>
> F. V. Hayden
>
> …If you wish any of the plants for yourself, take what you wish. If you think Prof. Torrey or Prof. Gray would like any of them, they are welcome to them.[12]

On the next day, April 6th, Hayden wrote another long letter, in this case to Baird. Besides containing another account of the "abortive" February-March trip to the Badlands, it includes an illuminating statement of the objectives Hayden hoped to achieve in the Upper Missouri region:

Fort Pierre, April 6th, 1855.

Prof. Baird,

My dear Sir,

I wrote you soon after my return from Council Bluffs. At that time I sent you three bottles of alcoholic specimens from the Yellow Stone which I hope you received. I now send the remainder by Col. V., consisting of one keg containing fishes, etc., one keg of embryos, etc., a bottle of things collected at Fort Union, etc. Among them all I hope you will find some things interesting. The collection is not so complete as I could have wished but for reasons which I will explain to you more fully when I see you.

As I told you in my last, by an arrangement with Col. V. all specimens except fossils go as a mutual present to you, though still Col. V. retains all the finest specimens of stuffed animals for himself. Yet I shall succeed in getting you some fine things by way of skulls, etc.... I will watch the spring birds narrowly. Please tell me if a collection of small birds in alcohol would be desirable to you. I find much trouble in skinning them, I suppose from want of experience. I will do the best I can. Birds are rare in this region for want of timber. All alcoholic specimens are also rare. There are so many mineral substances in the stream that but few fish are seen, and only a few [streams] are permanent, [most of them] drying up completely in the fall, and then freezing to the very rock in winter.

I made an abortive trip to Mauvaises Terres. Started 11th Feb., went as far as Sage Creek, a severe snow storm came upon us, and I turned back. Sent my team and men home, and took across to the Cheyenne trading houses alone from Pinoe's Spring about 35 miles, got lost, and spent three days in the terrible ravines of [the] Cheyenne with nothing to eat. Snow 18 inches deep. I then took a trip to the Black Hills with Mr. Galpin, the bourgeois of Fort Pierre, had a fine time, for 10 days we had only a meal a day and for four days on our return we ate nothing but one prairie dog divided among three. I came near perishing. I was so weak when I got to the Fort 22nd March that I could hardly sit on my horse. After getting at the Fort, we all ate so imprudently that we were taken very sick. In part I have not got over it yet. They tell me that now I have become partially naturalized in the country. I have had some bad times, but still if I can add my mite to the sum of human knowledge I shall not complain. It is very strange that at the base of the Black Hills a couple of men cannot kill game enough for their support but it is true.

I shall start again for the Bad Lands in a few days, with one man and a boy, three horses & a cart, to be absent six weeks. Though I go very light handed and almost unprotected, I hope to accomplish something.

It has been my intention until recently to return with the steamboat in July, but Col. V. is quite anxious to have me go up the Yellow Stone again and assist him in writing out his report. I am also anxious to go and review the labors of the past summer. For want of knowledge and experience I left many things incomplete, which now makes me feel unpleasantly, and as I shall return by September I think it best for me to go. However, I would like to ask your advice. Money is no object to me, provided I can make my researches as complete as possible, and reliable. I have several objects in view which I have partially accomplished and can finish in one more trip:

1st, a map of every stream and important locality from mouth of Platte to sources of Yellow Stone & Mo., with Indian names in the languages of Indians that live in the different portions, with explanations. Also a comparative view of all the languages on Mo., with small grammar, and vocabularies of each for the American Ethnological Society, which I have already about half completed and can finish it, I think, in my leisure moments next summer.

2nd, a review of my geological sections and collecting of alcoholic specimens.

This is a good deal and perhaps too much, but I shall try to accomplish it. I fear it is a fault with me, trying to do too much, and perhaps I had better do less and better, but I see so much to learn that it makes me greedy. I do wish you would make any suggestions and give me any advice you may think I need. I may err in many things, but certainly not intentionally, and my confidence in you and your friendship for me is such that I hope you will not hesitate. I have already derived much benefit from your letters, and I regret much that I could not have heard from you during the past year. Your last letter which I have received bears the date of June 3rd, 1854. But it is no fault of yours, for no express has come up since last September…

All the Fur Company feel a high regard for you. Mr. Galpin, the present bourgeois of Fort Pierre, has treated me like a brother. I owe him a great debt of gratitude. He is a pretty good practical geologist and Science may expect much from him. He expects to go below with Mr. Culbertson. I have requested him to call on you. At any rate I wish you would see him.

I have some wants now to tell you of. Since all the collections except alcoholic specimens go to you, please pay any freight that may be on them. Col. V. will not be willing to. The only expense will be from St. Louis to Washington. I wish you would purchase and send me by steamboat two kegs of alcohol, and one ream of plant paper. Dr. Engelmann will know what kind. Also some more tools for skinning animals and birds, material for nets, line bags, labels, etc. In all these things I shall be deficient by the time the steamboat comes up. If it is best for me to return by steamboat I shall not need them. Anything you may send to me if put on [the] load [of the] steamboat, care of Col. Vaughan, will come to me free of charge. I should like very much to get all that may have been written about this country since my absence. I presume Prof. Hall has written something, and I suppose Dr. Evans' report is now published, which I am very anxious to see. It will be a grand thing, and please [do] not fail to send it me by boat. Perhaps Dr. Leidy has published something; I wish it too. From all these works I shall get many suggestions which will make me examine more minutely. I will try to pay whatever they may cost in some way. Is anyone coming up into this country from below this season?

All my fossils will be stored very safely in St. Louis until I return. Col. V. will call on you at W. He has one tooth which I gave him. It came from White River. It is a large black tooth, and I can see nothing in Dr. Leidy's memoir that resembles it. Perhaps a new species. If so please let Dr. Leidy indicate it.

I have a large number more of them. Dr. Leidy has written very kindly to me. Please remember me to him kindly. I shall secure him some fine specimens for description. My friend Mr. Meek and myself will describe the shells.

I doubtless annoy you sometime with my long and frequent letters, but I do not know how to say less.

I shall ever remain most kindly and gratefully yours,

F. V. Hayden[13]

Although Hayden referred to the trip made in February and March 1855 as "abortive" and a "failure" because it netted him no large collections, this journey and the others made during his early years in the West were far from fruitless. As he penetrated the little known or unexplored country in divergent directions, whether by boat, on horseback, or afoot, he was observant always of the geographical features traversed, and gained an intimate acquaintance with Nebraska Territory which later proved invaluable to himself and others. Meanwhile, of course, he became experienced in living, working, and surviving in the wilderness. When finally expeditions were sent out into Nebraska Territory "under government," he was sought out for his familiarity with the country, which was unique and was regarded as virtually indispensable. Through the explorations of 1853-1855 he came to know the Badlands and other sections of Nebraska Territory of notable scientific importance better than any other scientist-explorer; in many instances he was the first scientist to visit and report on such regions. For example, when he reached the Black Hills in March, 1855, he became the first geologist ever to examine this geological feature, which since has become classic and world famous. Writing to Meek he observed, merely, "Visited the Black Hills, determined their geological structure," but when called on a year later to describe the regions traversed on this trip and others, he could write vividly and accurately from first-hand observations, remembered or recorded in his notes and journals. The following passages are illustrative:

> At Council Bluffs the river bottom becomes gradually wider, and extends without interruption to Floyd's Bluff, reaching its maximum width near the Little Sioux. At Floyd's Bluff the hills approach the river, and are cross-sectioned by it, and the wide river bottom appears on the opposite side. Here bluffs, as they are usually called, extend along the river about eight miles, to the mouth of the Big Sioux river. At this point the magnificent Vermilion prairie commences, and continues with very little interruption to Dorion's Hills, a distance of over 60 miles. All these bottoms are covered with a most luxuriant growth of grass and flowering plants, often rising to the height of 8 or 10 feet, effectually concealing a horse and his rider...
>
> About one hundred and fifty miles below Fort Pierre, the black clay bed of the Cretaceous system commences covering the whole country as with a pall, so dark and gloomy is the aspect. This bed extends along the Missouri to the mouth of Cannon Ball river, and over the region drained by the sub-systems of the Teton, Shyenne, Moreau, and a portion of Grand Rivers. Wherever it

prevails, the soil, if there be any, is exceedingly sterile, producing very scanty vegetation, and sometimes considerable areas present a bare, blackened appearance, without trace of tree or shrub. In the vicinity of the Great Bend below Fort Pierre, these "arid argillaceous hills" are extremely sterile, being much denuded, and presenting a burnt and blackened appearance, so that they have earned for themselves the appellation of "Cotes Brules" by the Canadian voyageurs. These hills are covered with fragments of selenitic crystals, or gypsum, which glisten in the sun's rays like gems, and from this fact are sometimes called the "Shining Hills." Above the mouth of the Cannon Ball river, the sands and clays of the Tertiary system commence, and the general appearance of the country is much improved; the upland is covered with a fine growth of grass, and more timber is seen along the stream.

About fifty miles below White river, is the first cedar island, a favorite wooding place for steamboats in their ascent of the river. This island is one of three which are destined to be of much interest to the settlers along the Missouri. They have an area of several hundred acres, and are covered with a dense growth, so crowded that the largest trees are not more than two feet in diameter. A large quantity of fine timber might be selected from these islands with advantage to the remainder. They are also the resort of myriads of birds and larger animals, which gather to these places to rear their young undisturbed.

The country in the vicinity of Fort Pierre, and north towards the Black Hills, though mostly sterile or entirely barren, contains some plateaus that are covered with a fine growth of grass. The region through which the main body of the Shyenne passes is sterile in the extreme. It is cut up by temporary streams into ravines and gullies to a frightful extent, and barren black hills, sometimes producing a few stunted cedars, meet the eye on every side. Between the two forks of the Shyenne, and in the vicinity of Bear Butte, a portion of the Black Hills, is quite a large area, which seems to have been levelled by denudation, and greets the eye of the traveller with joy, who has for some time previously seen nothing but the sterile country before mentioned. The streams which flow into the Shyenne at this point have their origin in perpetual springs in the Black Hills, and contain an abundance of fish, and swarms of beaver live along their banks. They are skirted with excellent timber of ash, oak, elm, and box-elder, and the level plains are covered with a fine carpet of grass which, in spring and early summer, is enamelled with myriads of flowers of the richest hues. On the 9th of March, 1855, I ascended Bear Butte, and on the south side, six hundred feet above the level prairie, I found a beautiful plant, (Anemone Patens,) just putting forth its blossom. The Indians call it the navel flower, for they say that when it blooms young spring is born. At that time the grass was springing up quite green, and herds of antelopes were quietly reposing upon the sunny sides of streams, like flocks of sheep. This is a portion of the country similar to White river valley, well adapted for grazing purposes, and capable of sustaining a tolerably dense population. The Black Hills which appear in the distance, and derive their name from their dark and gloomy appearance, contain an inexhaustible quantity of the finest timber, mostly pine, which will doubtless remain undisturbed for many years to come. I will, however, propose a plan for obtaining this timber, and render it useful to future settlers, though I do it with some hesitation, lest it may seem visionary. The left fork of the Shyenne passes through the northern portion

> of the Black Hills, and even there is a considerable stream, from thirty to fifty yards wide. In the spring the river is much swollen, and the current exceedingly rapid, and the timber, if cut and hauled to the banks of the river, might be floated down into the Missouri with considerable safety and ease.[14]

To resume our narrative: before leaving Fort Pierre for still another attempt to collect in the Badlands, Hayden affixed his signature to a letter to Baird, which is in the handwriting of Vaughan and restates the terms of agreement between Hayden and Vaughan:

> Fort Pierre, April 9th, 1855.
>
> Prof. Baird,
>
> Dear Sir,
>
> From the deep interest you have taken in the success of our enterprise from its commencement we feel it a pleasure to tender you as a mutual present all alcoholic specimens, skins of mammals, birds, plants, etc., which we have as yet secured or may secure until the close of our enterprise. We hope that among the specimens we have already collected that you may find some that will interest you, and we believe they will be of more value to you than any one else. If you will furnish us with two kegs of alcohol of five gallons each and any other aperatus [*sic*] that you may think necessary, we will take the utmost pains to make our future collections as complete as possible. Anything you may place upon the American Fur Co. boat to the care of A. J. Vaughan Ind. Agt will be conveyed up free of charge.
>
> We remain very respectfully,
>
> Yours,
>
> A. J. Vaughan, Ind. Agt
>
> F. V. Hayden
>
> The alcohol and paper for the plants which I previously furnished is free of charge. I hope to see you in Washington as I have much that is new to communicate pertaining to this truly remarkable region.
>
> Very respectfully
>
> Alfred J. Vaughan[15]

That same day, Hayden began his journey. The year was now so far advanced that inclement weather was no longer a problem, but for that very reason the Indians were again at large, and their belligerence posed a threat, one so serious that the trip had to be cut short. Compelled to return to Fort Pierre, Hayden gave expression to his disappointment in his next letter to Baird. Nevertheless, he still would not admit final defeat:

[Fort Pierre, about mid-April, 1855.]

Dear Prof. Baird,

I am disappointed again. I started on the 9th of April to Bad Lands with high hopes. Got as far as Grindstone Hill. Met a trader coming in from Brules camp who told such a doleful tale that my guide became scared and would go no farther with me. The Bad Lands are now full of the Platte Brules who say that they will rob every white man they meet on the Laramie road. They are determined that there shall be no communication between Fort Pierre, etc. and Fort Laramie. Everything seems to work against me. It seems as though I could do nothing. It will take me two years to accomplish the work of one under the present state of affairs. I am not discouraged yet, neither is Col. V. I have made arrangements to stay here until fall. It will take me until then to do what I wish on account of the troubles from rascally Indians.

I wrote you informing you that we had agreed to send you all specimens except fossils. I was glad to succeed in that. You will get a letter with Col. V.'s views to that effect. I would like to suggest (not order) respecting their disposal, — 1st, I hope you will take for yourself such specimens as may please you, these for Smithsonian Institution. I would like to have Prof. Kirtland obtain some skulls, fishes, etc. if he would like them. He has always been very kind to me and encouraged me when I most needed it. I hope before I return to get specimens for all my friends who have encouraged me.

I should think by your letters that you expected much to be done. Do not look for much. I labor under great disadvantages. I have to lose at least half of my time and much of that is without profit. I cannot do as I will — and could if I had means. In respect to fishes I may not disappoint you much. I have just got the hang of catching them, so that you may expect a fine lot from many localities when the steamboat comes down. If I stay at Fort Union any length of time this summer I will try to get to Red River, but I think it impossible…

I wish 2 kegs alcohol, one for Fort Pierre & the other for Fort Union. 2 reams of paper, size of plant paper. Labels. Line bags (I can make them perhaps), a pair of scissors like those you sent me; those were stolen from me. Some fine mustard seed shot — I am going to pay more attention to birds. Also, more of those vocabularies — Schoolcraft's vocabularies and questions… I wish very much Lyell's Principles of Geology. Mr. Galpin has promised to get it — and perhaps if you find it convenient to send me his latest edition I would find it of great value…

I feel very badly that I cannot get to Bad Lands, but Mr. Galpin says he will fix me up yet. Perhaps I shall have to give it up until the government sends troops up here and wipes out two or three hundred of them ("rascally Indians"). The traders are most all leaving the country, they have got so scared.

I wish I could hear from you. Not a word as yet since June 3rd [1854]. We know nothing about [what] is going on below. Col. V. is waiting for Mr. Culbertson who has not got here from Fort Union yet.

I would like some cotton, also about a peck of small boxes such as match boxes, tin mustard boxes, many kinds of tin boxes holding about half pint or pint. I find them very useful for recent shells and delicate fossils, and soils, etc. One keg of alcohol I wish at Fort Pierre for a trip planned in fall which will contribute much to Science. I am not now at liberty to state what it is.

I shall send a part of my plants to Dr. Engelmann, between 500 & 600 species. All these I can prepare for you if desirable. I have collected quite a number of fine species this spring.

Have you ever seen a white skunk? Mr. Galpin has obtained a skin from the Indians which is perfectly white all over — an anomaly in nature, I suppose. I have requested him to send it to you. I am collecting birds and fishes as fast as I have opportunity.

Pardon me for writing so long letters.

I am sincerely and truly yours,

F. V. Hayden[16]

A few weeks later, on May 7th, the indefatigable Hayden was off again on another and all-out attempt to reach the collecting grounds in the Badlands. This time luck was with him and he was successful. A letter that he wrote to Baird after his return to the fort includes a brief account of the trip, and is one of the most interesting documents from the period of Hayden's early years in the West. Hayden dated this letter May 10th but this must have been a slip; probably the date actually was June 10th:

Fort Pierre, May 10th [June 10th], 1855.

Dear Prof. Baird,

I have at last succeeded in my wishes in making a collection in Mauvaises Terres. I cannot tell you how much anxiety, trouble, and disappointment an attempt to visit that place has cost me. It seemed as if all Nature, animate and inanimate, was against me to prevent the success of my efforts, and in this third attempt I think I have been tolerably successful, though it was the most severe trip I ever made.

I started from the Fort under unpleasant circumstances. All said it was doubtful about my returning, but now was the time or never I thought. I had already been several months waiting an opportunity & making ineffectual attempts, so I resolved to go even if I had a collision with the Indians. An Indian at the Fort, who was starving, accompanied me as guide and spy, and I had two carts, one man & a boy 15 years old, all the men that could be spared, so you see I was about alone.

We took a new route, passed through the heart of the Brule country. Starting from the Fort, we struck White River about 50 miles above its mouth, passed then along the valley and crossed it above the Forks, passed along the dividing ridge between L'Eau qui Court & White River, thence to a point near the Porcupine just below where the Platte road crosses White River, to the heart of Bear Creek, a most interesting country I can assure you and which I shall be delighted to describe to you if I ever see you. This journey took us 17 days. I then worked hard 5 days gleaning in the old fields of Dr. Evans, Mr. Meek & myself of two years ago and returned to the Fort. I have my collection now safely in the Fort. Mr. Hodgkiss says that I might go ninety-nine times again and not escape for if we had met even 10 or 12 lodges [of Indians] we would have had trouble. As it was we saw only three lodges. They were in the Sand Hills, hunting buffalo, but a day or two before we left we saw their signal fires on White River, which showed that they had returned. They are now threatening to burn the Fort, but they cannot do anything I think, and when the troops come they will all be quiet as lambs.

O, if I could only work without fear, with ample time and means, with the influence that money or government could give me! It seems as though I could do something. As it is, I must pass over the ground hastily, gather up the most obvious things, as hastily return, for fear of Indians and for fear of expenses. I do not believe that the deposit has yet yielded up half its treasures and it can only be made to do so by the most patient and minute examination. I am confident that the different orders of Mammalia are well represented somewhere and that numbers of new species may yet be found. I find that the small patches of Bad Land are now the most interesting.

I hope I can add some new species which I think will please Dr. Leidy much. 1st is a Rodent, somewhat like a beaver I should think. 2^{d}, two or three species of small Rodents, as mice or squirrels. 3^{d}, a small skull, for which I know no analogous being. 4th, a few teeth of an animal allied to a horse. Col. V. has one of them with him which I requested him to hand to you. I am not aware that the Order Rodentia has yet been found here. I therefore ask, why should we not look for representatives from all the different orders? But it must be done by a faithful examination on the hands & knees, for I find in this way a third and even a fourth examination of a locality most fruitful.

I also found one splendid skull of M. primaevus, and a skull of one of the most ferocious looking animals I have ever seen. I think it must be allied to Hyaena or Tiger, being nearly of that size. Its canines are not perfect but must have been two or three inches in length. I found it on White River nearly opposite Pinoe's Spring. Among the large quantities of bones, teeth, and jaws I have, some things of interest may be found. I long to show them to you, and if you find it of interest I shall be paid for my labor.

Of fishes I made a pretty good collection for you from strange localities, three species of turtles, some serpents, and one lizard from Sand Hills. Those things most interesting to you I have the poorest facilities for getting and are the hardest to preserve. I will, however, do the best I can. If I stay up here another season, I shall go to the Forts above, where I can do more in that line in one month than in a year at Fort Pierre.

An express arrived from below, a few days ago, bringing three letters from you. They gave me the greatest pleasure. There must be some [others] on the way somewhere. I hope you have received those alcoholic specimens. Dr. E. is rather slow in sending them. I have much anxiety about them for fear some may never reach you. The three bottles I spoke to you of must contain some fine things. I trust these are the ones Dr. E. has. I also sent a keg to you by steamboat last summer. Did you receive them? Col. V. took along some which I hope you will get, though I fear they will not be well preserved. I also sent Dr. E. about 600 species of plants. I hope with the collection I made for Prof. Hall to make a Flora of Upper Mo. tolerably complete.

I wish to stay in the country until winter certain, and review the past year, and then I feel confident that something can be done to enable me to return and examine the country thoroughly. I ask no reward for past labors and only enough to support me for future. Only give me means to act. I am treated here with respect, yet I am well aware that here a man without money is a bore and with money a great gentleman let him do what he will, and my industry only makes me a greater bore. I have tried to make myself useful in every way. I have taught the children of Mr. Galpin & Hodgkiss to read, carried express for them. Yet I know I am a burden because they can make no money out of me. Mr. Culbertson is not so. He has very kindly invited me to spend the next season at the upper forts, and seems to feel a good deal of interest in my success.

With our whole collection I think a good deal of a show might be made at Washington. Would it not be better to have the whole brought to W.? If you think so, please suggest the thing to Col. V. in your next letter.

If a line of posts are established along the river, I can operate safely and thoroughly and without doubt the intelligent portion of the officers would encourage me. I will not hesitate to introduce myself to Capt. Van Vliet. I hope you have seen him or written him a line about it, so that I may not seem too abrupt.

I received a very kind letter from Dr. Leidy. He asks, "For whom are you now collecting and how are you especially engaged?" You can answer him better than I can. He also says he will see what can be done in the Academy. Please confer with him about it if you think best. I feel a great regard for Dr. L. and I think his kindness of heart is only equaled by his genius. I only hope I may be able to find some specimens in his line that will interest him. Is his forthcoming work to be published by the Smithsonian Institution? I shall send him a small box of my specimens in hopes he may find something to add to it. It will come to you first.

I am now preparing my collections of past year for steamboat, and I find it will weigh several tons, but I think there will be no trouble in getting it to St. Louis. The collections I am able to get, except fossils, are yours. Col. V. has a large quantity of stuffed animals, but I think hereafter you will get the rest. If I could spend a few days with you it would be well. As it is, my ignorance how to prepare these specimens causes me to lose about half or more. But I think I can manage better though.

If I do not go down this summer, I hope I shall get a good supply of alcohol. I will make good use of it now. As to smaller mammals, if I go above I can surely get all you wish, but in this region it is difficult. The Indians are too numerous and too mean. I hope I may go to Blackfeet.[17]

The Indians say the steamboats are coming up the river below us. I hope soon to have many letters from you.

I remain ever most truly and gratefully yours,

F. V. Hayden[18]

Through a fortunate circumstance, early in 1856, when Hayden had returned east and was in Washington, he was asked to contribute an account of "the Dacota country," which by this time he knew well, to a publication of the War Department. On relatively short notice he prepared this account and in it he incorporated what he termed "a brief digest of my journal made on a tour to the Bad Lands, in the spring of 1855." The report was published at Washington in 1856 as an appendix to a Senate Executive Document.[19]

As will become apparent in later pages, the publication to which Hayden thus became a contributor is highly important in the literature of western exploration. Hayden's part of it is significant for additional reasons. It was the first report to be published under his own name. The "brief digest" it includes is an excellent account of the successful trip which climaxed his attempts to reach the Badlands in 1855, one that in certain aspects is far more detailed than that contained in his letter to Baird, and in its leisurely narrative style it differs from the staccato writing characteristic of most of Hayden's letters. This is the most extensive excerpt from Hayden's journal, in anything like verbatim form, which got into print before those records disappeared.

Thus the digest of Hayden's journal merits a place in this volume. It is reprinted in the next chapter.

14

Spring Tour to the Badlands, 1855

This chapter is taken from Hayden's "Geological Notes, etc. on Nebraska," which forms a part of Appendix E in Lt. G. K. Warren's report.[1] Hayden's contribution is introduced by the following letter of transmittal:

Washington, D. C.,

March 15, 1856.

Sir: I herewith present a brief sketch of the geological and physical features of the region of the upper Missouri, with some notes on its soil, vegetation, animal life, etc.

Very respectfully, your obedient servant,

F. V. Hayden

To Lieut. G. K. Warren,

United States Topographical Engineers.

As is suggested by this letter, Hayden's part of the report includes not only the "brief digest" of his journal, herewith reprinted, and the descriptive paragraphs that are quoted above in Chapter 13 (pages 191-193), but also interesting observations on other subjects such as the hydrography, climate, plants, animals, and Indians of the Upper Missouri region:

> I started from Fort Pierre May 7, 1855 with an Indian as guide, one voyageur and a boy, with several horses and two carts, for the purpose of making a collection of mammalian and chelonian fossils in that remarkable cemetery [the Badlands]. The weather was warm, and sun shining brightly. Most of the usual spring birds had come, and the ravines and water courses were green with grass and foliage of trees, and many flowers were in bloom. The first

day's travel is usually quite short, not more than six or seven miles. We passed the steep, barren, Argillaceous hills that surround Fort Pierre, and encamped on the Teton or Bad river of the Sioux. We made about six miles, course southwest. The region drained by this river is quite sterile, the soil much impregnated with what is usually called Epsom salts in this country. In the dry portion of the summer and autumn the stream has no running water, and the bed and sides of the banks become incrusted with this salt; at a distance they seem to be covered with snow. The water standing in pools possess very active purgative qualities. This substance occurs very abundantly throughout the prairie region, and its medical effect is very similar to that of Epsom salts. Sometimes a very large area of many acres is covered with it, several inches in thickness. I do not know that any analysis has been made of it. It is very fine, and when pure, white as snow, and much the consistency of flour.

The following day we pursued the same direction, over a country having much the same sterile appearance, destitute of water, and camped in an immense depression, called by the traders "Big Hole." In it is the bed of a little stream that empties into the Teton river, but at this time it was nearly dry; the water standing in pools and very unpleasant to the taste. Our wood for culinary purposes was small willows, and our horses picked a scanty meal in the moist ravines. All the upland prairie was parched with drought.

The following day, May 9, we pursued nearly the same direction, and through a similar country, destitute of wood and water, the vegetation parched by drought. No game was to be seen but a few antelopes, which were so wild that it was impossible to kill them. About three o'clock we reached the head of Medicine creek, a tributary of the Missouri. Our distance was about fifteen miles. We found here excellent grass for our animals, fine clear water, and an abundance of wood. This stream is fed by springs, contains an abundance of fish and well skirted with timber, and much vegetation grows along its banks. Many species of flowers were in bloom, rendering the contrast quite pleasing.

The following day we continued the same course until we came to the White river valley, and the upland prairie presented the same parched appearance. Approaching White river, the country is somewhat rugged, and the hills that bordered the valley had the usual arid sterile aspect, but we soon passed over into the beautiful valley which was at this time clothed with fresh green grass, from ten to fifteen inches in height and the timber skirting the river was in full foliage. Myriads of flowers enamelled the prairies and honey bees were industriously at work by thousands among the Leguminous blossoms. I was not aware that this little inhabitant of civilization had wandered so far into the western wilds, until he appeared here, and I would infer from the number I saw that they must inhabit some of the numerous dry trees along the river. After striking the river we passed up along the broad level to the forks, and camped in a most delightful place. We were in want of nothing, either for ourselves or animals, and of course all in high spirits. The soil of these bottoms is very fertile, composed of the calcareous and aluminous marls of the Tertiary basin through which the upper portion of this river flows, and the clays of the Cretaceous system which forms the hills, and is the basis formation throughout the valley.

The following day, May 11, we passed up the valley four or five miles above the forks, for the purpose of fording the river. The left fork is the main stream, and at this time was much swollen and the current quite rapid. The bottom is so full of quicksands and other impediments that we found it very difficult to discover a crossing place. Finally, after a most diligent search of several hours wading about in the stream, we ventured to make the attempt. We first unloaded everything in our carts and carried it to the opposite side, a distance of one hundred and fifty yards, and then letting our carts down the steep bank with great care, we passed safely over on the other side with no other trouble than the miring of our horses in one or two instances. After landing safely on the other side, we ascended the high hills which border the river, and camped in a beautiful ravine, from which flowed a fine spring, surrounded with an abundance of wood. The whole country here has much the appearance of the limestone region near the mouth of the Platte, and I noticed several species of shrubs that are found no higher up on the Missouri than Council Bluffs. Our distance that day was not more than eight miles. The water of White river is very peculiar, containing a large quantity of calcareous and aluminous matter held in suspension, so that it has much the appearance of milk. When allowed to stand for a short time, or whenever it is found in pools, a thick scum may be seen upon the surface very much of the appearance and consistency of rich cream; removing this, and the thinner portion is of a much lighter color, like milk. It is very astringent to the taste, and its medical effect on the traveller is quite the reverse of the water previously used. This river has been navigated by traders with small Mackinac and bull boats for over one hundred miles from its mouth. We found plenty of antelopes near our camping place. Our Indian killed two fine ones, which formed a most acceptable addition to our larder.

On the following day we passed over a fine rolling upland prairie on the northern slope of the dividing ridge between White river and Running Water. After descending two steep hills which border the river bottoms, the prairie slopes gently towards the south for a distance of fifteen or twenty miles to a high ridge which is visible in the distance and covered with pines. Our course was about due west. On our right the lofty denuded ridges of the "Bad Lands" appear, while scattered all over the valley are patches of white earth, monuments of the Tertiary basin left after denudation. The whole valley gives unmistakable evidence of having been levelled to its present condition by the action of denudation. The day was quite fine. We camped at night on Bad Land creek, and before reaching it we passed through considerable bad ground, which rendered our travelling quite hazardous. They seemed quite appropriately named. There are many things of the deepest interest in this valley, and as yet it is almost *terra incognita*. All the streams on the right side of White river flow from perpetual springs in the distant ridge, and even in the dryest season they are full of cool crystal water, their sides fringed with many varieties of water plants, and their valleys clothed with abundant grass and flowers. Contrasted with most of the country on the upper Missouri, the White river valley is a paradise, and the Indians consider it one of the choice spots of earth. Indeed, when supplied with an abundance of meat, they always resort here and spend their time in amusements or in cultivating their numerous gardens which are scattered throughout the valley. The Brules and

Yanktons raise a considerable amount of corn and other vegetables, and with very little attention they get a tolerable crop. The greatest deficiency that will be felt here will be the scarcity of wood for fuel, and, like other portions of the upper Missouri country, no coal need be sought after.

May 13. — We again travelled due west across a fine rolling prairie. We met with large numbers of a peculiar species of cactus just coming into bloom, (*Mammaria nuttalii.*) It is more abundant here than anywhere else on the Upper Missouri. We camped at night on Grass Creek — another fine stream, similar to those before mentioned. At our left, in the distance, we could see a tall, naked white butte, called Eagle Nest butte, from the fact that an eagle has built her nest on the summit from time immemorial. Although it rises to the height of one hundred and fifty feet, with nearly perpendicular sides, its stratification is perfectly horizontal, presenting a more conspicuous example of the terrible denuding action which has been felt in the valley. On its summit I found several species of plants in bloom that I observed no where else on the route.

May 14. — We passed several fine streams, all having a uniform appearance, and the whole region approaching more closely the fertile portions of Kansas. We camped at night on a fine stream called "The creek where the Indians plant corn." Our Indian guide spent a day here planting a small patch of corn and potatoes. On our left, as far as we could see, was a fine rolling prairie clothed with vegetation, while on our right, the tall naked columns and domes of the "Bad Lands" loom up, assuming most unique and fantastic shapes. Our Indian guide led us along an old buffalo track, a very rugged road, a fact of which we were convinced by having our carts upset and the contents distributed over the ground. The Indians say, however, that in travelling over a country with which they are unacquainted they always follow a buffalo trail, for this animal always selects the most practicable route for his road. At the mouth of this creek is the most beautiful exhibition of the Bad Lands I have ever seen.

On the 18th we camped near a fine spring, which was but a little distance from it, and we resolved to stop a day or two and explore it. Our animals needed rest, and here was an abundance of good grass and water. After partaking of a delicious dinner of antelope meat, I started out, accompanied by my voyageur, and ascending an elevation which was above the bad ground, looked down upon one of the grandest views I ever beheld. The denuded area was nearly square in form, and the immense flat concretions that projected out from the sides of the perpendicular walls in regular seams, and at about equal distances above each other, resembled some vast theatre; indeed, it reminded me of what I had imagined of the amphitheatre of Rome, only nature works upon a far grander scale than man. We climbed with great difficulty down the steep sides, following the main channel of the little stream, and after much winding through this labyrinthian sepulchre, we came to an open plateau covered with fine grass, and in the centre a beautiful grove of cedars, and through the whole a stream of milky water wound its way to White river, about five miles distant. All around us were bare, naked, whitened walls, with now and then a conical pyramid standing alone. We felt very much as though we were in a sepulchre, and, indeed, we were in a cemetery of a pre-Adamite age, for all around us at the base of these walls and pyramids were heads and tails, and fragments of the same, of species which are not known to exist at the present day. We spent that day and the following exploring the cemetery, which the denuding power of water had laid open for our inspection, and many fine specimens rewarded our labors.

On the morning of the 21st, we started for White river, on which we camped. We were about to leave behind us good grass and water, for in the direction we were going we could expect nothing but a desert and salt water. The bottoms were, as usual, quite good, but the hills that bordered it were terribly denuded, showing us that we were still in "Bad Lands." The following morning we crossed the river with some difficulty, on account of the sinking of the horses into the quick sands, and then wound our way among the sand hills in a direction nearly parallel with the river. These sand hills sustain very little vegetation, have often very nearly perpendicular sides without stratification, and are probably formed by the mud.[2] On our right, the main body of the Bad Lands appeared with its numerous labyrinthic passages, illustrating most finely the denuding power of the water. Many rare plants were observed here, but few animals. One species of lizard was quite common, but very agile in its movements. We succeeded, after many attempts, in catching one of them, by throwing a blanket over it suddenly. Antelopes were more abundant than other kinds of game, and we were able to supply ourselves with meat. Occasionally a big horn was added, which made a very agreeable variety. The meat of the female has much the flavor of mutton. On the following night, we camped on a tributary of White river; wood was very scarce, and the water of the poorest quality. All the tributaries of the northern side of White river are of the same character. The next day we were desirous of getting to Bear creek, the usual camping place of visitors to the Bad Lands. Our road was exceedingly rugged, winding among the terrible canyons and gullies. We were not able to get through without again upsetting our carts, and were obliged to pack our baggage a considerable distance. At last we struck the main road between Fort Pierre and Fort Laramie, and then our troubles were at an end. We had followed an old buffalo trail seventy-five miles through a country never before travelled by carts. We had been seventeen days reaching Bear creek, whereas, if we had followed the usual route, it would have required but six days. But I had the pleasure of exploring a large area of most interesting country, never traversed before but by Indians. The locality at Bear creek has yielded the large and valuable collections which have already been brought to the States by Mr. A. Culbertson, Captain Van Vliet, Dr. Evans, and others, and furnished the material for the magnificent work of Professor Leidy [1853], "The Ancient Fauna of Nebraska," published by the Smithsonian Institution. We spent five days at this locality, and with the mammalian remains already collected in other places, our carts were loaded to their utmost. I noticed one peculiarity in the atmosphere at this place — the frequent showers that annoyed us very much, for they kept the marly earth around us in a disagreeable condition. In passing out of the Bad Lands our carts were much impeded by the mud, but when we ascended upon the plateau above, the storms ceased, and thirty miles east of that point we could see no indications of its having rained for several weeks. From Bear creek we made an ascent of two hundred feet to a broad plateau, extending, with little interruption, to the head of Teton river, a distance of about fifty miles. The formation upon which this plateau rests is a yellow arenaceous bed, much more fertile than the black clay bed in the region of Fort Pierre. It is covered with a good growth of grass, and affords a fine pasture ground for numerous herds of antelopes.

On the night of the 30th, we camped at Sage creek, so noted among voyageurs and travellers, for the purgative qualities of its water. I stopped here several days in the summer of 1853, with my friend, F. B. Meek, esquire, and both ourselves and horses experienced its weakening effect. This creek rises in the Bad Lands, and taking a northeast course, empties into the Shyenne. It is somewhat noted for the abundance of fine Cretaceous fossils that are found along its banks. For about four miles east of Sage creek the road is very rugged, until we again ascend to the table land. Twelve miles distant is Bull creek, also a tributary of the Shyenne, and takes its rise in the Bad Lands.

The dividing ridge between the Shyenne and White rivers, forms the most elevated portion of the Bad Lands. Fifteen miles east of Bull's creek, is Pinon's spring, a noted camping place among voyageurs, as being one of the very few good springs of water on the road between Fort Pierre and Fort Laramie. It is the head of the Teton river, and is so called in the Sioux language. It runs in an opposite direction to the creek before mentioned, for about four miles, then passing through the eastern portion of the Bad Lands, becomes the Teton, takes a due east course, and empties into the Missouri four miles below Fort Pierre. Near this spring is an elevated ridge composed of white aluminous and calcareous marl, containing slabs and concretions fully charged with fossil shells of the genera *Lymnea, Planorbis, Physa, Helix*, etc., also seeds of Chara and fish remains. Some of the shells have already been described by Drs. Evans and Shumard. It is a curious fact, that though these shells are of extinct species, living shells of the same genera, and plants of living Chara, are now existing in the stream that runs at the base of the ridge. The remainder of our route we crossed the heads of the other tributaries of the same stream, with very little wood and poor water, and at this time the earth was so parched that very little good grass could be obtained for our animals. On June 6 we arrived at Fort Pierre safely, having been absent thirty days.

15

To Fort Benton and Return

For Hayden, the eventful winter of 1854-1855, with so much of both accomplishment and frustration, had become a memory. The spring of 1855 too had run its course, and as we have seen it brought him an opportunity to run the gauntlet of the hostile Sioux, reach the Badlands, and return to Fort Pierre with his life and collections intact.

The oncoming of spring meant the end of a kind of hibernation for most of the occupants of Fort Pierre. It brought surcease from cold and snow, escape from the cramped quarters of the post, revival of more active pursuits of one kind or another, and a welcome change to more varied and palatable food.

Spring marked the end of isolation since overland expresses could now get through to the posts on the Upper Missouri. Also, as ice thawed from the rivers, navigation was resumed and steamboats again labored upstream laden with passengers, freight, and mail. With communications restored, regular monthly mail service was provided. Hayden at long last got news from his friends in the states by the early May express, the first mail to reach him since the previous September, and later by other express deliveries and by boat. Probably it was the first express that included the following letter from Meek, one which had been five months in reaching Hayden:

Albany, New York, Jan'y 6th, 1855.

My dear Sir,

Your two letters, one from Belleview and the other from Council Bluffs, both reached me some time since, and I owe you an apology for not answering them sooner. In as much, however, as I have a very good one to offer, I hope you will take no offense at my seeming neglect of an old friend.

Your first letter arrived at Columbia during my absence, and I did not see it for some time after, when I returned from the survey of Cooper & Moniteau Counties. The other came to hand when I was working with all haste to have my report ready to hand in by a given time; which made it necessary for me to write every moment my eyes would bear, and my engagements have compelled me to start immediately on as soon as my report was completed. Consequently I have postponed my answer until the present late day.

Before leaving Columbia, I had a conversation with Prof. Swallow in relation to a position for you in the Geological Survey of Missouri. He appears to have changed his mind in regard to doing much with the Botanical department at present, and says he is not authorized to employ any more assistants in any other part of the work at this time.

On my arrival here I found a letter for me from Dr. Parry[1] dated at Davenport, Iowa, in which he states they are getting up a survey in that State and that he is confident of receiving the appointment of State Geologist, in which case he offers me the place of Palaeontologist. If I had confidence enough in my own ability to perform the duties of so important a position, I cannot see how much an arrangement could be made consistent with my present arrangements here and in Missouri. In addition to this, I am now satisfied my eyes are too weak to undertake such a work. I will write Dr. P. in a few days and recommend you for a position as an assistant in the Survey, should it be ordered. It would also perhaps be well for you to write him on the subject. I presume he does not wish to have this matter and his expectations in regard to it, known beyond the circle of his confidential friends at present.

In relation to describing your collection from the Upper Missouri, I scarcely know what to say. If any arrangement can be made by which we could be together here or elsewhere I would certainly be delighted to join you in the work. I cannot see, however, how it would be possible for me to go to Philadelphia to do it, as I am compelled to spend four months here and the remaining eight months of each year in Missouri. Time may, however, yet turn up something by which some arrangement may be made, so that I can join you in it.

Dr. Shumard failed to obtain the appointment of State Geologist of Ky. and is still holding his position as Palaeontologist of the Missouri Survey, the duties of which he discharges with great ability. I sent him your letter.

I have much to say but I must close as I have some three or four other letters to write, and I am also hurried with the drawings which have accumulated during my absence. I only arrived here day before yesterday.

Yours in haste,

F. B. Meek[2]

When Hayden returned from the Badlands, on June 6th, he found Fort Pierre rife with rumors, excitement in the air. It was common knowledge, by now, that for Fort Pierre one era had ended and another had already begun: the fort had been sold by the Chouteau Trading Company to the War Department. A new order of things had come to pass, one which obviously would affect everyone connected in any way with the post.

The altered status of "historic old Fort Pierre" (which had been founded in 1832) was but one indication of far-reaching changes taking place in the West. Fundamentally these changes involved little that was new; they were simply further manifestations of the continuing course of westward U.S. expansion. Changes were now proceeding at an accelerated pace which was precipitating new and grave problems. The reasons for this are not far to seek and may be briefly explained.

During the early 1850s, the so-called "depredations by the Sioux" had become frequent. In reality, the "Sioux wars" (which were to flare with intermittent violence for four decades) had already begun. The situation was perhaps inevitable, for occasions for friction between Sioux and whites were multiplying as ever larger numbers of emigrants streamed through the Sioux hunting grounds, over the Laramie Road and other trails, and as settlers thronged to the frontiers in Kansas and Nebraska territories. Thus the Grattan massacre of August 19, 1854, an account of which has been quoted from Vaughan's annual report for that year, was no isolated incident.

The Grattan massacre was notable, however, because it hastened a more critical phase of the conflict with the Sioux. Its violent character aroused widespread demand for retaliation against the guilty parties, and for more effective protection of western emigrants and settlers. Following this incident, the situation was deemed intolerable, and in the War Department plans were laid for dealing decisively with it. Implementation of these plans wrought great changes in the conditions under which all who were in the West pursued their respective activities, not least so Hayden.

Orders from the Secretary of War dated March 22, 1855 designated Brevet Brigadier-General William S. Harney[3] as commander of a force of about one thousand troops assigned to the field to operate against the Sioux. It was assumed that the theater of operations for the impending campaign would be limited on the south by the Platte River, on the northeast by the Missouri River, and on the northwest by the Black Hills. The region thus defined was about 90,000 square miles in area. Strength of the Sioux was estimated at about 7,000 warriors. A decisive engagement with the whole band was held preferable to several engagements with small, scattered bands. To accomplish the desired end, it was decided to establish three depots for the collection of troops and supplies: Fort Kearney, Fort Laramie, and a third depot to be situated on the Upper Missouri at or in the general vicinity of Fort Pierre.

On March 23, 1855 the quartermaster at St. Louis was directed "to obtain the most reliable information possible as to the suitableness of Fort Pierre… for a depot of supplies."[4] John B. Sarpy, the active partner at St. Louis of the firm of Pierre Chouteau, Jr. & Co., was approached and gave a quite candid opinion concerning Fort Pierre, namely, that it was not suited to serve as a depot of supplies for any considerable force. He painted a rather dismal picture of it: "The fort itself is small, and is located in the mauvais terre [*sic*], where for hundreds of miles there is no grass that can be made into hay; no good ground for corn and fodder, and no fuel within twenty miles."[5] He had to admit, however, that for the purposes of the contemplated operations there was no better alternative location. The more favorable views of others prevailed, and presently it was decided that the United States should purchase from the firm, for $45,000, "the trading establishment on the Missouri River, near the mouth of the Little Missouri River, called and known as 'Fort Pierre,' together with all buildings within and around the pickets of said fort, and all the lumber and other materials in and around it, 'as well as' the island in the vicinity."[6] The sale of Fort Pierre took place on April 14th and possession was to be given on or before June 1, 1855.

Orders for the movement of General Harney's forces had been issued on March 23rd. Some troops were instructed to rendezvous at Fort Kearney, and others at Fort Laramie; the rest were to proceed up the Missouri by boat to the depot in the Fort Pierre vicinity. The depot last-mentioned was to be the principal one; here two-thirds of the supplies were to be accumulated, and accommodations were to be provided for four companies of cavalry and six of infantry. The river traffic which took place that summer was unique in character and was extraordinarily heavy.

Movement of troops up the Missouri began in June and was complicated by various difficulties, especially obtaining vessels of proper draft for carrying the many soldiers and their supplies, and hiring pilots and crews of sufficient experience and competence to cope with the navigational problems of the river. Consequently the steamboat carrying the first consignment of troops to Fort Pierre, the *Arabia*, did not reach its destination until July 7th. In the ensuing weeks of July and August, most of the other boats managed to get to the fort, each with its quota of troops.[7]

Colonel Vaughan embarked from St. Louis on June 6th (the day on which Hayden got back to Fort Pierre from the Badlands) as a passenger on the *St. Mary*, a boat operated by Pierre Chouteau, Jr. & Co.[8] With Vaughan were Alexander Culbertson and also Vaughan's superior, Colonel Alfred Cumming, Superintendent of Indian Affairs from St. Louis. Cumming was planning to spend the season with Vaughan on his annual tour to the tribes of the Upper Missouri. Furthermore, he had a special mission of exceptional importance in that he was senior emissary sent by the United States government to negotiate a treaty with the "Blackfoot Nation" (that is, the Piegans, Bloods, Northern Blackfeet, and their allies the Gros Ventres).[9]

The first stop mentioned by Vaughan in his *Annual Report for 1855* was made at Handy's Point, which he identified as being on the northeast side of the Missouri River, thirty miles above the mouth of the L'Eau qui Court. Here was situated the principal village of the Yancton band of the Sioux; nearby were the fields of corn, beans, pumpkins, and other crops which these Indians cultivated in abundance.

Vaughan duly reported the cordial reception given him by the Yancton. After this brief halt, the *St. Mary* continued its course and reached Fort Pierre on June 26th. Vaughan had anticipated meeting most of the Sioux bands here, but found only the Two Kettles band and some of the Upper Yanctonnais (27 lodges). Though these Indians accepted their annuities, they did so without tokens of gratitude. "On the contrary," reported Vaughan, "they seemed dissatisfied and evinced a disposition to complain." He added, "I cannot account for their conduct in any other way than [that] as the troops were to take possession of Fort Pierre they would be deprived of a place at which they loafed and begged the greater part of the year, only going out to hunt when actual necessity drove them."[10]

When the *St. Mary* docked at Fort Pierre on June 26th, Hayden came on board and he accompanied Vaughan and Cumming on the remainder of the journey by steamboat, to Fort Union. Thus his travels began anew.

At the time of Hayden's departure from Fort Pierre, troops were being expected but had not yet begun to arrive. In this connection it may be noted, parenthetically, that at about the same time that the *St. Mary* left St. Louis another fur company boat started up the Missouri.[11] Among the troops which it carried was Lieutenant Gouverneur Kemble Warren, a brilliant young soldier who would play a major role in the mapping of the West and who would have much to do with F. V. Hayden. Warren was enroute to the West for the first time, under orders to report to General William H. Harney as chief Topographical Officer. From the correspondence, it is known that Spencer F. Baird of the Smithsonian Institution had written to Hayden about Warren, and it appears probable that he had also communicated with Warren about Hayden in an effort to unite the endeavors of the two men. Warren's trip up the river brought him into "Hayden's country," and had his boat made good time he would probably have met Hayden at Fort Pierre. However, Warren's boat was much delayed by "daily groundings, the constant difficulties with sandbars, and heavy, baffling winds," and it did not reach Fort Pierre until July 16th. By that time, Hayden was gone and already at work far up the Missouri. One may conjecture that if Warren had reached Fort Pierre in time to catch Hayden there, the latter might have made a last-minute change of plans and have then and there cast his lot with Warren. If so, Hayden would probably have gone overland with Warren to Fort Kearney in August in order to join General Harney; and he would have been present when Harney's forces clashed with Sioux on September 3rd. (This bloody one-sided engagement, at Blue Water Creek near the present city of Ogalalla, Nebraska, cost the lives of eighty-six Indians and five soldiers.)[12] As it was, the trails of Warren and Hayden did not converge at all that summer and their first actual meeting occurred the following winter, in February 1856 at Washington, D. C.

The *St. Mary* reached Fort Union on July 11th, "thirty-five days out from St. Louis, notwithstanding the river was lower than ever has been known before by the oldest settler in the country."[13] Their experiences between Fort Pierre to Fort Union were, judging from Vaughan's report, rather routine. About 100 miles above Fort Pierre, Vaughan distributed goods to twelve lodges of the Yanctonnais, who were making their first attempt to form a permanent village and cultivate the soil. Wrote Vaughan, "I am sorry to say that the great drought in that region of their country was such that all kinds of vegetation presented but a very languishing appearance." The Yanctonnais received their presents "with evident indication of satisfaction."

At Fort Clark, reached on July 5th, the Arickarees were all assembled on the bank of the river and "greeted the arrival of the steamer with firing of guns and shouts." The village of these Indians comprised 60 dirt lodges. Vaughan met the chiefs and principal men and after the usual salutations and hand shaking invited them all aboard the steamboat to a feast. There a long talk ensued "principally relating to the depredations and murders on their people by the Sioux bands, their inability to cope with them in numbers, as well as their destitute condition." Vaughan states, "They rejoiced when they heard that their Great Father had sent soldiers in the country to chastise those who had violated their treaty stipulations, and protect those who have and are disposed to observe them." Vaughan was gratified by the reception these Indians gave him, by their talk and deportment, and by their prosperous condition, for despite the drought they were raising "a superabundance of corn and vegetables, the large surplus of which they dispose of to the neighboring tribes and traders."

On the evening of the same day, Vaughan distributed goods to the Mandans. "Though their portion was small, and their wants many, not a word of complaint was heard from them." It gave Vaughan gratification to report that the remnant of this tribe was increasing and now numbered "21 lodges, built after the manner of the Arickarees… making an aggregate of 252 souls." Like many another, Vaughan was intrigued by the question of the origin of the Mandans and the significance of their peculiarities, such as "the diversity of their complexion of skin, eyes, and the great length of their hair." It was his opinion that "there is no ground for considering their origin different from other tribes around them." He added, "They are truly a domestic people, remain at their home, and till the ground with success."

The *St. Mary* reached Fort Berthold and the village of the Gros Ventres on July 7th. Vaughan wrote approvingly of the Gros Ventres, who numbered 40 lodges aggregating 760 souls. "The friendship of this tribe to the whites is well known, and they have strictly regarded their treaty stipulations. The large fields of corn and vegetables which covered the bottom lands… showed their industry and great desire to obtain the means of improving their conditions." The pastoral life of these people was now darkened by an ominous cloud cast by the belligerent Sioux. "Their complaints against the Sioux for stealing their horses and murdering their people were any thing but pleasant. It pleased them much to hear that soldiers had arrived in the country, and [they] consoled themselves that soon a better state of things would be made to exist."

At Fort Union only a few of the Assinaboins were on hand; but two days after the boat's arrival, 350 lodges presented their appearance, and the Indians encamped on the level prairie west of the fort. The Assinaboins were well supplied with provisions, and "having a heavy trade to make" (presumably with the fur company) they remained quietly at camp until a few days later, when — as a result of Vaughan's having dispatched a messenger — the rest of the tribe arrived. There followed a general session at which Vaughan, before distributing the goods, gave the Assinaboins a lengthy address. Talk and presents were both received with satisfaction and gratitude. Vaughan wrote that "since the Laramie treaty these Assinaboins are the best Indians on the continent. It was truly gratifying for me to learn from the lips of the principal of this tribe that not a single instance of murder, robbery, or other depredation has been committed by them, either on the neighboring tribes or on the whites. This is the more remarkable, as before the treaty... they were foremost in the van of thieves and robbers, always at war, pillaging from whomsoever they met, and were very annoying at all trading posts." The Assinaboins, who numbered 420 lodges aggregating 3,360 souls, had a good health record too during the past year. "I am happy to say," wrote Vaughan, "they are prosperous and happy, always cheerful, for they have plenty to subsist on, as game of all kinds is still abundant in their country."

In reporting on the Assinaboin, Vaughan took occasion to salute "our energetic and efficient superintendent, Colonel A. Cumming, whose presence among the various tribes we met with made a favorable impression, and who's [*sic*] counselling will result in future good."

The arrival of the *St. Mary* at Fort Union on July 11th was noted in the journal of James H. Chambers, an employee of the fur company who was at the fort at this time. His entry for the day stated:

> Wed. 11 — we had the pleasure of greeting the arrival of the Steamer *St Mary* quite a number of gentlemen arrived among which were A. Culbertson Esqr . Col A. Cumming, Supt. of Indian Affairs Col A. J. Vaughan Ind. Agt & Major Hatch of the Blackfoot Agency & several other gentlemen.[14]

The "several other gentlemen" included F. V. Hayden, and we will now turn to Hayden's correspondence for information about the events of the summer, particularly his own activities.

So far as known, Hayden's first letter of this trip was one written to Baird from on board the *St. Mary* the day before the boat docked at Fort Union. Preparations for unloading the remaining cargo were already under way, as the *St. Mary* would not go beyond Fort Union because of the low water; and to escape the uproar surrounding him Hayden sought refuge in the brig. Here, in comparative quiet, he wrote the following letter:

Fort Union, July 10th, 1855.

Dear Prof. Baird,

I received several letters from you by steamboat which gave me much satisfaction. Yet I do not believe I have all you have written me. One has [a] date of Nov. 17th, another May 10th, & 4 in all. I was very sorry that I could not have received them sooner. I might, perhaps, have done more for you in collecting small mammals.

You will see by my letter that I am on my way up again toward the mountains. Mr Culbertson very kindly invited me to go to Fort Benton, and seemed desirous of aiding me all in his power. I think he merits in a great degree the respect of scientific men for the interest he has felt, and still feels, for Science. I shall always remember him with much gratitude.

I did not see Lieut. Warren. None of the officers had arrived when the boat left Fort Pierre though they were supposed to be on the way. I shall write a line to him, requesting him to consider me attached to his corps if he wishes me. Otherwise, I will avail myself of all the facilities I can get. Col. Cumming, Sup. Ind. Aff's [at] St. Louis, is on the boat, on his way to make a treaty with the Blackfeet. He offered of his own accord to give me a letter to commanding officer at Fort Pierre which should assure me all the facilities I might need. I begin to think I shall have a chance to do something. Hitherto I have been obliged to act on so economical scale that I have lost about half my labors, especially in skinning. I have a large box & a barrel of skulls & skins at Fort Pierre for you. All my specimens are there. Whether they will all be destroyed by the soldiers or not I can not tell. Mr. Charles P. Chouteau, a son of P. C., Jr., has charge of the steamboat this year and proposed to take all my specimens down and store them in St. Louis, free of expense, as aid to Science, but I fear the water will be too low on their return…

I have 18 boxes of geological specimens, 1 box plants, box & barrel of skins, etc. and a number of loose horns & heads. I secured a pair of Prairie Foxes (V. Velox), the most interesting little fellows I ever saw, which I was intending to get to you alive, so that you might study their habits, but Col. V. gave them to Mr. Chouteau. Should you ever go to St. Louis, he would be pleased to have a call from you. Write me if any other animals would be useful enough to you alive to pay the trouble of getting them to you, and I will make a great effort another Spring.

In the box of skins you will find 2 bear, from Fort Sarpy, 3 skins black tailed fauns from one mother in Bad Lands, 4 antelopes from 2 mothers — one pair had not yet been born, the other pair were one day old.

I had no chance about Fort Pierre to get skins, in fact [was able] to do but very little, and I am glad to get away. One of the hunters going up to Fort Benton is a good friend of mine, and he will favor me much, so that I hope to send you a fine set of skins in the fall. At any rate, I shall always do all I can, but I am never able to accomplish half what I think I ought to.

I received from Dr. E. 2 kegs alcohol, one ream paper. That is my complete outfit for this year. No books, reports, nothing else. I would give half my collection for Lyell's Principles of Geology. I have not a single work on geology now.

I suppose that Gov. Stevens, Dr. E[vans], etc. have explored the Blackfeet country so thoroughly that there is little left to be done. I feel desirous of going up, however, inasmuch as I will have a chance to go up the Yellow Stone next season. I have found a great many friends who encourage me and are disposed to assist me, full as many as I deserve. I think the way is now open to stay in the country as long as it will be worth a while. Of course I have some dark hours when it seems as though all my hopes were about to fail, but I expect them in my dependent condition. But if I can accomplish my object all will then be well.

Dr. Engelmann sends me a lable [*sic*] which he has had printed. "Flora Nebraskana." This pleases me. Will not the Smithsonian Inst. publish a list of Flora and Fauna of this country when I get through? I have gathered some insects. Will you have them named and a list of them made out? I have some fishes and I intend to get you a full collection from Fort Union to the mountains. Among them you may find some new species. Will you describe them as fast as they come to you? I do not hope to find many things that are new to you, and I am not surprised that there is nothing new as yet in my collections, but I may aid in determining the geographical distribution of animals.

I hope you will write to any of the officers you may think best, introducing me to their favorable notice. It will help me much in my operations.

Mr. Kipp is on board, on his way to take charge of Fort Union. He is the only white man that ever spoke the Mandan language. He has promised to give it to me. Mr. C. also has promised to aid me. But these men look at the surface of all things, and I am obliged to ply them so closely sometimes as to annoy them. I propose to make my notes reliable. I think I can do something in this line.

I hope now that the facilities for getting specimens to you will be better. They ought not to delay a moment anywhere. I hope Dr. E. will forward them as soon as he gets them. Col. V. has made a bargain with Mr. Chouteau to take all our geological specimens down [and] store them. [Mr. Chouteau] Takes one half of Col. V.'s share, pays one half of the expenses already incurred and also one half that may be incurred until next fall. Col. Vaughan also proposes to send you some specimens to dispose of, which I hope may be of some value to him.

There has been so much confusion and hurry on the boat, and now instructions [?] of going 300 miles above Fort Union, so now I suppose the boat returns immediately. I write in a brig, perhaps my suite[?]. Excuse poor writing.

Ever gratefully yours,

F. V. Hayden[15]

On July 11th, Hayden wrote another letter, in this case to George Engelmann. Whether at the time of writing he was still on the *St. Mary* or had already moved into Fort Union is not stated:

Fort Union, July 11th, 1855.

Dear Dr. Engelmann,

I received a long, excellent letter from you by steamboat, the only one since September last. I was glad to hear from you, though I wish I could hear oftener. I have written you several long letters which I fear never reached you.

I am glad if you have found any Cacti that interest you. I have some more, also some notes which you will receive by steamboat. A box of plants are at Fort Pierre. You will see by the outside of the bundle what ones were collected this year. The others I would like you to preserve. Have you discovered any new or rare species among them? Please let me know in your next. I hope yet to make the "flora Nebraskana" very complete. I am pleased with the labels. I hope you will make a catalogue as full as possible.

You doubtless received a box of plants from Col. V. in the Spring. You will get some things by the boat which I hope you will forward immediately to Prof. Baird. I wish to direct your attention to a small box, marked Prof. Baird. It contains all my discoveries in Bad Lands, 4 or 5 new species. If they should get lost in any way I am broke. I hope you will get them and see them sent to Prof. Baird in the safest manner possible.

I will send you my notes, which owing to my limited time may puzzle you some though I hope you can make use of them. I attempted to describe the plants peculiar to this country. It is not done in botanical language, but [*sic*] as I have no books. You may make something out of them. Among them you will find some Cacti. Since I came up the Mo. from Fort Pierre I have found several curious forms of the flat leaved Cactus, perhaps a distinct species. I have preserved some of the forms for you.

I am now on my way to Fort Benton. From thence I expect to go to the mountains. I shall there look for other Cacti forms and those parasitic plants you spoke of to me. I am told they are abundant.

I am glad to hear that your little son is becoming a naturalist so young. He shall have some fine skulls in the fall. I would have sent him some by steamboat but I had them all packed away to send to Prof. B....

I hope to collect every thing that grows, whether common in the States or not. Is it not a matter of interest to determine the geographical distribution of even the commonest plants?...

Since writing the above, Mr. C. P. Chouteau has agreed to take the whole collection down [and] store it away carefully till I come. He is connected in the matter and will take an interest...

I have just sent you (July 12th) some loose bundles of plants, and a box of Cacti with a few other things in the box. There is a large paper of recent shells, which you will see, all very fine, found in driftwood below Fort Berthold. Save all carefully. If you can I hope you will go immediately to see to the things personally.

I will write again by the Opposition Boat when it returns.

I remain ever yours,

F. V. Hayden[16]

Two other letters written by Hayden at Fort Union are preserved. The earlier of these, addressed to Baird, was dated July 12th; the later, to Meek, was dated July 17th:

Fort Union, July 12th, 1855.

Dear Prof. Baird,

The boat is detained one night longer than expected, and I get time to write a few words more. I send you what alcoholic specimens I have on hand. I have nothing but an open keg, which I hope Dr. E. will fix for you.

In it is a jar with, 1st, at bottom, a specimen given you by Mr. Culbertson. He says it is an anomalous buffalo calf's head. 2d, serpents from Fort Pierre collected in May. One greenish snake with a string tied to its neck is from White River valley. The others were all found about Fort Pierre. At the top is a bag of insects. A few were collected on my trip through White River valley, most of them about Fort Pierre, in June. Passing along White River I made a small collection of fishes which I will try to label so that you may distinguish them.

No. 1, (dark skin labels) [from] creek above the Pouka by or Pouka creek, a tributary of White River. May 12.

No. 3, Eagle nest pass creek, about 40 miles west. May 14th.

No. 4, from a branch of Pouka Creek. May 12th. Also, 3 fishes from head of Indian creek, and all the loose fishes in the keg. May 8th.

No. 5, a long bag with various specimens from Chekie (?), a branch of Titon or Little Mo. on road to Bad Lands. June 4th.

...Col. Cumming proposes to attach me to his corps, so as to give me a position giving me a nominal salary, pay all my expenses. He intends to go by way of Red River to St. Peters, from there to St. Louis. I wish I could have your advice in this matter, whether it would be best for me to do so. I am anxious to confine my labors to the Missouri country, and at the same time go to St. Peters.

Mr. C. P. Chouteau wishes to take a share in the labors of Col. V. and myself. Col. V. proposed to let him have one half of his share if he would take the collection to St. Louis and one half [of] expenses already incurred. He accepted. I have reserved the right of description.

A small box, I hope, will come to you. You will find various things in it but at the bottom 4 tin boxes with curious jaws and teeth, which must be handled by no one but yourself, also some match boxes, a small turtle... Please examine them if you choose and send them to Dr. Leidy for description. The other things, please preserve them. I wish they could all be in a better condition, but I have no materials now and I can get no kegs. Should I go with Col. C. I will fish Red River for you. Do not fail to let me know the condition of my specimens when you get them. Most of them are full, but it is difficult to get them down safely. Please send me one or two more copper cans. They are very fine. Let me know if there is anything special you can direct my attention to... write to Lieut. Warren and tell (him) I will be with him in the Winter or Spring.

As soon as the confusion is over and I feel a little better I will write again, as I shall be at Fort Union several days.

I ever remain most sincerely and gratefully yours,

F. V. Hayden[17]

Hayden wrote to Meek as follows:

Fort Union, July 17th, 1855.

Dear Friend Meek,

The steamboat has just arrived at this place and all is bustle and confusion as you saw it at Fort Pierre. It was confidently expected that the boat would go 300 or 400 miles above this point, but low water prevents, and I am behind hand with my matters.

I am now at this place on my way to Fort Benton in the Blackfoot country, 780 miles from this place by water. I expect to have a grand time of it, and in addition, I will be at the treaty with the Blackfeet.

My specimens thus far collected will go down in the steamboat in the care of Mr. C. P. Chouteau, son of P. Chouteau, who is somewhat interested in them, and I am under some obligations to return in the winter and arrange and describe what I have collected. Can you not manage to be in St. Louis December and January, when I will meet you, open the boxes, and select such [specimens] as we wish to figure and describe? Let me know what you can do. Can you not spend 2 or 3 months next winter in St. Louis or any other place you may designate? Be sure to do it. We can make something fine, I can assure you. I would like to have you see my collection and you would not then hesitate a moment to make any sacrifice to work it up.

I told you in my last of making a successful trip to Bad Lands. I passed completely through the Valley of White River, and returned with two cart loads of bones & turtles, etc. Some new discoveries. I found several quarts of a Helix identical or similar to the one now in the possession of Prof. Hall...

Let me know somewhat of your calculations, also what has been done below in way of publications or getting up surveys. Do you not know that I am ignorant of every thing? The steamboat did not bring up any newspapers, not even a magazine. I never knew such a time. Don't fail to write me a long letter.

You have heard of the changes that have occurred here? Fort Pierre is sold for $45,000. I have received a proposal to become attached to Lieut. Warren of Topog. Engineers who is now at Fort Pierre. Col. Cumming, Superintendent of Indian affairs, goes to Blackfeet, from them to Pembina & St. Peters [rivers] to St. Louis. He proposes to attach me to his corps. This would give me a fine chance. What to do about it I do not know. Perhaps it would be best for the purpose of connecting the geology of Mo. with Miss. and determining more fully the boundaries of Cretaceous & Tertiary.

Do not fail to make all your calculations to join me next winter. Get Prof. Hall to excuse you one month. I could write much. Give my kind regards to Prof. Swallow and Shumard. I remain ever your friend,

F. V. Hayden[18]

Hayden set off from Fort Union on Tuesday, July 17th, bound for Fort Benton and the mountains, his "farthest west" to date. On July 17th Chambers noted in his journal: "Boats left for F. B. [Fort Benton] Mess Kipp & Dawson in charge Major Hatch Doct Haydon & several other passengers."[19] The boats mentioned in this entry were keelboats and mackinaws, which were to relay the trade goods brought to Fort Union by steamboat on to destinations farther up the Missouri, including Fort Benton. To reach Fort Benton, they would have to be cordelled against the current a distance of nearly 800 miles. A hundred men were employed for this job, that of walking along the banks of the river and pulling the boats by means of long tow ropes.

Though Chambers mentioned Hayden as being a passenger, and later Hayden himself wrote Baird that he "went on keelboats to Ft. Benton," it is unlikely that he remained continuously with the flotilla of boats carrying the trade goods. These heavily laden boats made such slow progress, and met with so many delays, that it was the middle of October before they reached the mouth of the Judith, and they did not get to Fort Benton until November 5th. In all probability Hayden left these boats somewhere along the way and advanced to Fort Benton with the aid of other, faster boats or by overland travel, or both. There is no record in the Fort Benton journal to indicate with certainty when he got to this place, though he may have been in a party which reached the fort on August 17th. An entry for that day mentions the arrival of "A Culbertson & family with Col. Cummings [*sic*] & two other gentlemen with B. Champaigne & several Indians."[20]

For a period of almost five months, that is, from July 12th to December 8th, we have no letters from Hayden. It is not likely that he wrote any during this period, inasmuch as he was in the wilderness and, most of the time, on the move. On this exploratory venture he was out of touch with his scientific friends for a longer time than on any of his previous trips. Vaughan, writing from Fort Clark on September 29th to Charles E. Galpin at Fort Pierre, observed, "Your friend Doctor Hayden took a trip up to Fort Benton, a region of country he has been long anxious to visit." Although little is known about Hayden's journey, one cannot doubt that he made the fullest possible use of the opportunities it gave him and that he did have the "grand time of it" he had anticipated.

All of the region along the Missouri upriver from Fort Union was new to Hayden; geologically, it was practically unstudied country. As in the summer before, when he accompanied the keel boat up the Yellowstone to Fort Sarpy, doubtless he tried to see as much as possible of the country bordering the Missouri on both sides, traveling freely at his own pace and on his own schedule, and returning to the boats or to camps only when convenient or necessary. He saw, for the first time, the areas of badlands which Culbertson and other men of the "fur brigade" had described to him; these he afterward referred to in letters and publications as the "badlands of the Judith," the "Mauvaises Terres of the Blackfoot country," or the "badlands of the Upper Missouri." As they proved to be features developed in Cretaceous formations, rather than in Tertiary beds like those of the White River Badlands previously explored, they yielded new assemblages of fossils. The collections he made were probably the first of real significance ever obtained from the region.

Though some information may be had from letters he wrote after the trip in December and January, and from a few other sources, we are left in the dark on many subjects. What route did he follow? Was he alone or did he have companions? How did he manage with his collections? Did he employ carts, as on his earlier trips into the White River Badlands? Did he, again, enjoy his usual immunity from serious trouble with the Indians?

As regards the last question, it would seem remarkable that anyone could ramble through this region in the summer of 1855 making scientific collections and notes without sooner or later encountering difficulty with Indians. The accounts in such sources as Vaughan's Annual Report, Chambers' journal, and the Fort Benton Journal give a graphic picture of the violence committed by Sioux warriors who ranged far and wide throughout the Upper Missouri region. Vaughan, for instance, in attempting to reascend the Yellowstone with annuities for the Crows, had scarcely begun his trip before Sioux warriors attacked his men and drove them back to Fort Union. The prospects of his party getting through to Fort Sarpy were so poor that he abandoned the attempt. The record for the summer of 1855 is full of such incidents, involving murder, plundering, and worse. Entries in the Fort Benton Journal from October 4, 1854 to September 27, 1855 list no fewer than forty-eight outward bound or returning war parties which stopped by that fort, and there must have been many other war expeditions during that period which did not come near the fort.[21]

In view of Hayden's phenomenal capacity for covering ground, and his avowed intention of going to Fort Benton and the mountains, it seems probable that he both visited the fort and reached some of the neighboring ranges, outliers of the main chain of the Rockies such as the Judith Mountains, the Bearpaw Mountains, or the Little Rocky Mountains. There is also a possibility that he may have followed the Missouri upstream from Fort Benton toward or into the Three Forks region where the Missouri has its origin in the confluence of the Gallatin, Madison, and Jefferson rivers. The speculation is intriguing, inasmuch as this would have been Hayden's introduction to a region which in later years held a special attraction for him: to it he returned many times, not only during the palmy era of the Territorial Surveys but also in his declining years, when he returned to the West in the vain hope of regaining his health, even as late as the summer of 1886, the year before his death.

Whatever the course of his rambles among the basins and mountains of the Fort Benton region, by the middle of October Hayden, along with many others, turned up at the rendezvous site appointed for the memorable council with the "Blackfoot Nation." Originally planned for 1854, the Council had been rescheduled for early August 1855; but due to the slow progress of the boats hauling treaty goods up the Missouri, it had to be postponed until these boats finally got to the mouth of the Judith River. Thus it did not take place until the middle of October. Had the Council been held in August, as planned, about 12,000 Indians would have attended, but the delay caused many Indians to leave the vicinity, so that only about 3,500 were present. Even so, it was an extraordinary gathering. "For the first time in the history of the west the Indians east of the mountains and the Indians west of the mountains were to be bound by a treaty of amity. Warfare was to be outlawed, and the western Indians were to be permitted to hunt buffalo east of the mountains without fear of their ancient foes of the Blackfoot nation, who regarded the western Indians as poachers... Warriors who had previously met only on the field of battle, or had taken part in expeditions for plundering each other, forgot their past differences."[22] There were present fifty-nine chiefs, the leaders of an Indian population which occupied an area more than twice the size of New England. The commissioners representing the United States government were Alfred Cumming, Superintendent of Indian Affairs from St. Louis, and Isaac I. Stevens, Governor of Washington Territory. Cumming outranked Stevens, being designated the senior commissioner, but Stevens had been working for this treaty for two years and proved to be the dominant figure.

The rendezvous site was about a hundred miles below Fort Benton. "The council ground was a wide, level plain covered with a noble grove of huge cottonwoods. It was on the left bank of the Missouri, nearly opposite but below the mouth of the Judith. This stream was also bordered by broad bottoms, which were covered with large sage-brush, and fairly swarming with deer... lower down was the camp of the crew of men who had dragged the boats up the river. There were a hundred strong, mostly Germans, having many fine voices among them, and were fond of spending the evenings in singing. The effect of their grand choruses, pealing forth over the river and resounding among the lofty trees, was magnificent."[23]

Here, finally, on October 16th and 17th, was held "the most important intertribal gathering ever held on the northwestern plains."[24] Despite personal antagonism and bickering between Cumming and Stevens, two days of negotiations took place.[25] These were brought to a successful close late on October 17th, at which time a treaty was signed, by the commissioners on behalf of the United States, and by chiefs, head men, and delegates of the various tribes present. At 4. P.M. the council was declared concluded. Many months later the treaty was delivered to Washington, and in April 1856 it was approved by the Senate and by the President of the United States. In large measure it achieved its purposes, the most important of which was to bring peace among the tribes represented, and between them and their neighbors and the whites. The Blackfoot Treaty "marked the beginning of a new era of Indian history" and has been termed "a masterful stroke of diplomacy, because it brought about a great change in the history of inter-tribal relations, and placed the conduct of the tribes under the authority of the United States."[26]

Culbertson was one of the important figures at the Council, serving as interpreter for the Blackfeet. Vaughan and Hayden also assisted during the negotiations in subordinate roles. Vaughan acted as a government reporter; Hayden's duties are not known. However, the compensation which Hayden received for his services helped to relieve his financial needs, which by this time were acute. Andrew Dawson, a member of the fur company in charge of boats carrying goods up the

Missouri to the Council and to Fort Benton, wrote a letter on October 16th to the St. Louis office, headed "Judith River" (presumably the council grounds) and added this postscript: "I have this day drawn on you for One Hundred Dollars ($100.) favor of Dr. F. V. Hayden which you will oblige me by paying on presentation. And. Dawson."[27]

During the three days following the signing of the treaty (that is, October 18th-20th) goods were distributed to the Indians, coats and medals were presented to the chiefs, and the commissioners made speeches exhorting the Indians to keep their promises and restrain their young braves from further warfare. Then the two government groups, headed by Cumming and Stevens, departed from the council grounds, heading in opposite directions. Neither of the co-commissioners ever again returned to the Blackfoot country. In his journal Stevens noted, "We got through with the Blackfoot treaty, everything having succeeded to our entire satisfaction beyond our most sanguine expectations. The greatest delight and good will seemed to pervade the minds of all the Indians, and we left them at the mouth of the Judith on our way to Fort Benton, and thence to the waters of the Pacific, rejoiced that our labors had had such a consummation."[28]

Hayden states in his letters that he left the mouth of the Judith on October 22nd, that is, five days after the signing of the treaty. He accompanied Cumming and Culbertson on the long journey back to Fort Union and thence to Fort Pierre; Vaughan was probably in the group also, at least part of the way. Most if not all of this trip was made by boat; Hayden's letters give no hint of their taking a route by way of the Pembina, Red, and St. Peters Rivers, as was evidently the intention earlier. Presumably they came directly down the Missouri, traveling by boat. From Fort Pierre on they had to proceed by land as it was November by this time and the river was frozen. At Hermann, Missouri, roughing it was finally over; the final stage of the journey to St. Louis was made in luxury by "the cars."

The first specific information from Hayden himself about the journey up to Fort Benton and back is found in a brief letter to Meek dated December 8, 1855 from Sergeant's Bluff. Hayden wrote in answer to a letter from Meek, delivered to him at this place. This exchange of letters follows, with Meek's letter first:

> Salina County, Mo., Oct. 23, 1855.
>
> My dear Sir,
>
> It has been a long time since I had the pleasure of hearing from you, though I have written you several times. Last spring I inquired at the Fur Companies [*sic*] house in St. Louis if they expected to send any dispatches to the Upper Missouri before the return of their boat and was informed they would not. Being under the impression that you would come down on the return boat I did not think worth while to write you to the Upper Mo. country. When the boat did come down I was out in the back wood, and did not know for about two months whether you have returned or not or where you were. Not long since I was informed you are still in the Upper Missouri country and I intended to write you at once but have put it off until the present time.

I have been in Mo. since last June and am consequently not very well posted in regard to recent events in the east. Our 1st Missouri report is, or soon will be, ready for distribution. It forms a large octavo volume of some 430 pages, and is illustrated with maps, sections, diagrams, etc. together with several plates of scenery, new fossils, etc. I will send you a copy as soon as I can have one at my disposal.

The Legislature appropriated last winter 25,000 dollars to carry on the survey and it is still in progress. Prof. Swallow allows me to work most of the time in another part of the field and to report the results of my own work in my own name. Dr. Shumard has the south, Prof. Swallow the north, and I the middle portions of the state. The work is progressing rapidly, and I believe to the satisfaction of the people.

I am as usual to go back this winter and stay at Albany until 1st May. Prof. Hall's salary in N. Y. has been fixed at 2000 dollars per annum. He also holds the place of State geologist of Iowa for which he gets $1300 per annum for 3 months work of each year. He also works up the Palaeontology of Canada. I do not know what he gets for the latter. I believe Prof. Leidy has not yet published his work on the Bad Land fossils. Dr. Evans is still in Oregon and will not report for some time to come. The memoirs prepared by Prof. Hall and myself for publication are not yet out. I will send you copies as soon as they come to hand.

Dr. Newberry went to California last spring as geologist of one of the Rail Road surveys. He is to return this winter.

I hope you have made a splendid collection and that we can carry out our old plan in regard to them. Could it not be so arranged that you could bring them to Albany, N. Y., if you should return while I am there? I think I could bring about such an arrangement to the satisfaction of all parties. If you write me so that your letter would reach me between the 1st Jany. and 1st May next, direct to Albany, N. Y., after that time direct to Columbia, Mo.

I had many things to say but since I have commenced writing I cannot think of half [of] them.

Yours in haste,

F. B. Meek

P.S. I would advise you by all means to try to keep down any ill feelings between yourself and a certain scientific gentleman of the East, for whom I believe you have no very ardent attachment. I know of no one man who can do a young man wishing to follow scientific pursuits more harm or more good than he. I have not heard him say much, but I believe he thinks you do not like him.

F. B. M.[29]

Sergeant's Bluff, December 8th, 1855.

Friend Meek,

I have got this far on my way to the States, and write you in hopes that it may reach you at Columbia. At this place I got your letter of Oct. 23 desiring me if possible to bring my collections to Albany that we may describe them. That will please me, if no other arrangement can be made. I have just returned from the Blackfoot country where I have made very extensive collections. The Mauvaises Terres of Blackfoot country I have examined well and I think is scarcely less interesting than the Mauvaises Terres of White River. I have with me a set of the fossils which came down in one baggage wagon. I must have at least several hundred undescribed species. I am rejoiced that you still propose to join me. It will only be the first of a series of fine things which we will do, as the future will show. I can only offer you this as your just due, that your name stand first, for instance Meek & H., and I will present you a set such as you will select of all my collection. This is all I can do. I hope you will feel satisfied. I have many things to say to you. Do not leave St. Louis until I see you, as I wish you to be there when I open my collection. C. P. Chouteau took it down in boat. I have about a ton at Fort Pierre now. I shall write again from Council Bluffs. I am with Col. Cumming, Sup't Indian Affairs, and was at the Blackfoot Treaty. I will write more fully from Council Bluffs. Do not leave St. Louis until I see you. I shall be there I hope the 25th.

Yours most sincerely,

F. V. Hayden

I have written this letter with a stick.[30]

Hayden's hope of getting back to St. Louis by Christmas proved too optimistic. Having fallen behind schedule, and being uncertain that he could still reach Meek by letter, he next wrote to George Engelmann, mainly with a view to getting word to Meek:

St. Joseph's, December 21st, 1855.

Dear Dr. Engelmann,

I have got thus far on my way to St. Louis and this has been done with a good deal of difficulty. We all got frozen up at Fort Pierre and came from there by land. Have encountered much bad weather & bad roads. I hope to see you by the 1st of Jan. but possibly a little later. I have heard nothing from you since July. Am sorry that I could not have heard whether my collections got to St. Louis safely.

I am very anxious that Mr. Meek should not go on to Albany until I see him. I wish him to spend a few days with me in my collections. I wish you would impress upon him the necessity of waiting for me so that we can select such material as we wish to describe and figure. I hope you will not fail to see Mr. Meek, and I hope this letter will reach you before Mr. M. leaves for A.

I am well, but I need not say to you that I have had a hard time of it and many disappointments. I have, however, done the best I could in all things.

I will write little as I will see you soon.

Most truly yours,

F. V. Hayden

Please give the enclosed note to Mr. Meek when he comes.[31]

Further unforeseen difficulties made it necessary for Hayden and his companions to revise their timetable, as they had to remain in St. Joseph until several days after Christmas. The enforced delay gave Hayden the opportunity to write Spencer Baird a long letter which gives considerable information about his recent experiences and activities. It also sets forth his thinking, at the time, with respect to future plans. These, though uncertain, included a number of possibilities:

St. Joseph, Mo., Dec. 26th, 1855.

Dear Prof. Baird,

I have not heard a word from you since last July, neither have I had an opportunity of writing to you. I took the steamboat at Fort Pierre, went to Fort Union, thence on keelboats to Fort Benton, started from the mouth of Judith 22d October, froze up at Fort Pierre, and then came the rest of the distance by land to this point. I can assure you that I have had a severe time of it and many disappointments. We have encountered several severe storms. Near Fort Clark the Sioux stole all the fine mules which Col. Cumming & Mr. Culbertson had provided for our land travel in case we were overtaken by ice. This put them to much trouble, and there has been a continued series of troubles all the season.

I have, however, made something of a collection. At any rate I have done the best I could. When ever I have had an opportunity I have labored with all my might. But I have failed to get but very few of my specimens down. 2 small boxes were brought to Council Bluffs, my choicest and new species which I intended to use this winter, but I could not bring them [further]. I however packed my notes and a very few specimens on a mule from Council Bluffs to this place. I very much wish you could see my collection.

I have not heard a word from Dr. Engelmann concerning those things which I sent down in care of Mr. C. P. Chouteau in the summer. I only know that they left Fort Pierre safely, all but two or three boxes, one of which is a large bread box filled with choice skins which I expect will be wholly spoiled before you can get them. I have made a collection of skins this summer sufficient to fill two large boxes, namely some bear, black tailed deer, 2 male antelopes, 1 male Big Horn, heads of male & female with skins on them, over 20 prairie dog skins, some rabbits, etc., a few bird skins, etc., some skulls. The copper can is at Fort Pierre, full of fishes, etc. from the various little creeks that run into Mo. above Fort Union. I have with me 2 bottles, one of insects and the other miscellaneous, as bats, mice, squirrels, etc., a shrew like the one before sent down, etc. These I will send to you as soon as I get to St. Louis. I wish I could send you thousands when I send you one, and perhaps you will wonder why I have not done better. I can only say that I did all I could, and what I have done has cost me much labor.

I had a fine opportunity to explore thoroughly the Mauvaises Terres of the Upper Mo. I mean the Bad Lands of the Blackfoot country. Although in most respects entirely different, I think them scarcely less interesting than the Bad Lands of White River. As I hope to show you soon, I made a small, but I think it will prove valuable, collection of vertebrate remains, which I wish to pass through your hands to Dr. Leidy for examination.

Did you ever receive the small box that I sent down in the summer, which contained the specimens which I supposed valuable? It is strange that Dr. E. did not write me, inasmuch as there has been a monthly mail to Fort Pierre all summer. I am yet in ignorance of the fate of my collections. If all is safe, my geological collection will amount to about six tons collected from a thousand localities and illustrating every phase of the geology of the Upper Mo. as far as I have traveled. I have detailed sections from Council Bluffs to Fort Benton, and about 400 miles on the Yellow Stone, while I have followed up White River, Teton, Cheyenne, Moreau, etc. traversing the country in different directions. I hope something will result from it of value to Science. I feel very anxious to submit it to your inspection.

I wish to ask your advice on many points and I trust you will advise me as you may think best for the interest of Science and my interest. I have accumulated an immense body of notes on a good many subjects. I propose first, if it meets with your approbation, to write out a large octavo volume, say 300 or 400 pages, entitled, "Wandering in the Far West, or three years on the Mo. & its tributaries." The kind of life that I have led has given me a better chance to accumulate facts than any one could have in merely passing through the country. I have material enough for several volumes, and I think I can get enough for one that will prove of some interest.

I have paid much attention to the Indian languages. I do not know what has been done in that way, and perhaps others may have anticipated me, but I will tell you. I have obtained large vocabularies of the Gros Ventres of the Prairie, Gros Ventres of Mo., Assiniboin, the Arikara & the various dialects of the Dakota, pretty good grammars & dictionaries of the Blackfoot, Crow, & Mandan languages. The whole would make quite a volume.

And then my geological notes & description of fossils, of which I must have several hundred new species, will be something more. How can I work this material up? & when? Can you advise me? Mr. Meek wrote me a letter which I received at Sergeant's Bluffs expressing a desire to join me in describing fossils. I am rejoiced. You know I could not have a better associate or more able. I have written him requesting him to await at St. Louis until I come, so that we can look over my collection together and select such as we wish to figure & describe.

I ought to have been in St. Louis in Oct. or November, but I have been so much under the control of circumstances and misfortunes that I shall not get there until after the 1st of Jan. I came to this point on horse back, with Mr. Culbertson & Col. Cumming. Here Mr. Culbertson purchased a splendid span of horses & carriage for $550, and last evening one of them [the horses] was nearly ruined, so that we will be detained here some time.

What would you think of an attempt to get an appropriation for a Geological Survey of Kansas & Nebraska? My friends throughout this region have said much about it and if you think it would be successful I wish it might be done. If I could get an appropriation for a preliminary survey, I think I could make it so thorough that it would not need further search. I have been laboring thus far with my hands tied as it were, and I do long for the freedom which some sort of position would give me, when I can have means to work all the time and not be under the control of circumstances and humiliated by poverty. If [it] should not be advisable, I have a half a dozen openings to explore after my former fashion. All the officers up at Fort Pierre have offered to assist me, and Gen. Harney made this expression, "Anything I have is at his service." I have at this place had a request to go to Salt Lake and other places, etc. All this might be done with very little expense. Do write me as soon as you get this, that I may have a letter from you on arriving at St. Louis. Advise me what to do.

Mr. Culbertson wishes to be remembered to you kindly. He will be in Washington this winter. I feel so anxious to see you and have a long talk with you that I can hardly wait for the time to come. Could you not meet me in St. Louis and look over my collection?

I do not feel very well at this place though I hope to feel better when on my way.

I remain ever sincerely & gratefully yours,

F. V. Hayden[32]

The difficulties which had beset Hayden and his companions on their return trip were almost over; however, one more delay is mentioned in a letter Hayden wrote two days before he reached St. Louis. It is clear from this letter (addressed to Baird from Danville, Missouri) that the hardships of the journey had not dampened Hayden's enthusiasm for a Geological Survey of Nebraska and Kansas territories such as he had urged a year earlier. Doubtless this hope was one he had never lost sight of during the long months of his absence, and now having regained the settlements where he again found kindred minds among the military and civilian leaders of the territories, it was renewed and his ardor was greater than ever:

Danville, Jan. 9th, 1856.

My Dear Professor Baird,

I am obliged to stay over a day here, waiting for a team to take me to Hermon, where the cars will take me to St. Louis. I hope to be in St. Louis tomorrow night. Mr. Culbertson goes about 10 miles from this place to visit his son and I will get to St. Louis as soon as possible.

Some new developments have been made which I wish to tell you of and which I hope will result in good. As we passed through Platte City, Mr. C. & myself called on Gen. Atchison.[33] We were received with great kindness. I presented to him the idea of a Geological Survey of Nebraska & Kansas, and he took it up with enthusiasm, said I need not fear that an appropriation would be made [at] the present session, that it was important and must be done.

He says the Surveyor General tells him that one million of acres has already been surveyed in Kansas and nearly the same amount in Nebraska and will be ready for sale in the Spring, that there has been so many conflicting accounts of the resources of the Territories that it is time now that there should be something official that can be relied on.

He told me how the thing should be managed, will write to the Sec. of War who, he says, is an old school mate of his, and set him right, also to Sec. of Interior and any other friends he may think of who will feel an interest, detailing the reasons why there should be an immediate survey. Of course you know Gen. Atchison's influence. Mr. McClelland is an old acquaintance and townsman of Mr. Culbertson and to him I can gain ready access. My friends in Nebraska & Kansas tell me to "command him" at any time, and memorials might be presented to Congress from the Legislature & Councils of the Territories if necessary. But Gen. Atchison says there is no need of that, that the matter is only to be presented and an appropriation will be made without hesitation. I am very anxious to hear from you what you think. If you say push the thing through not a stone shall be left unturned to lay open the vast store houses of Natural History knowledge in these vast Territories. I long to work untrammeled and above board, and explore every hidden recess of this intensely interesting portion of Nature.

But then the question arises in my own mind, can I get the appointment? And am I fit and capable of filling such a position creditably, and would it conduce to the best interests of Science? I leave that for you to decide, and if you think it not best, I am content. I do not wish to presume at all, but I do think that if I can explore a wild country alone and without pecuniary means, I can certainly do so with the advantages such a position would give. I long for the position for no gain, only that I may be enabled to work with greater success and contribute more to the knowledge of this region in which I am especially interested. My experience in prairie life and my knowledge of the country, already acquired, would (it seems to me) enable me to accomplish much more than one not thus accustomed. I long very much to see you and talk with you.

I can hardly tell you what a terrible time we have had getting down. We have been traveling toward home ever since the 22^d Oct. and O how we have suffered from storms and cold. This will be the severest winter in the mountains that has been known for many years. Yesterday we rode 40 miles, the thermometer in the morning 24° below zero, and from 8° to 10° below during the day. I have been riding through the country with Mr. C. in a private conveyance. Mr. C. has been very kind to me and wished me to stick by him or I should have been in St. Louis by the 1st of Jan. But I have an excellent opportunity to get a general idea of the country through Iowa & Mo. without expenses. Besides I wish to investigate a point in geology, namely, the yellow marl hills which you will see noticed in Dr. Owen's Report. I think I can now solve the problem concerning them. They are the most conspicuous geological features in this region. I hope to submit a paper to you this winter on that subject, that may be of some value. I made some examinations at Glasgow which will aid me much.

Do not fail to write me at St. Louis as quickly as you can. Your last letter to me bears date of July, but I hope to find a good many in St. Louis. Col.

Cumming, Sup. of Indian aff's St. Louis, will be in Washington in a few days. I hope you will not fail to see him. You know he has been up to Fort Benton. He has aided me much and I think will give you a good account of me.

I wrote a long letter from St. Joseph which I hope you received. I told you I had a fine lot of vertebrate remains which I wish to send to you for Dr. Leidy to examine. They are mostly single teeth and vertebrae and will, I think, reveal a new feature in the geology of that country (Blackfoot country). I also said that I wished to prepare a narrative volume for publication. I think I can get it ready in two months. Mr. Denig[34] is now collecting and preparing material for a great work on the Indians of Upper Mo. for Culbertson. Mr. C. has with him manuscript for one volume, which I have requested him to show you at W. I think you will find more reliable information in it than in a score of other works on Indians. Mr. C. has agreed with me to write a physical description of the country, essays on Indian languages, etc., and revise the whole and see it through the press as joint author. It will be published a year from this winter, in two large octavo vols. well illustrated & under the auspices of Mr. Culbertson. You see I have a good deal to do before me but I will bring it all out right if my health is spared.

I saw Gov. Stevens and spent some time with him at the Council ground. I will send you some letters from him as soon as I get to St. Louis. I hope to see you soon in Washington.

As to money, I have some. I have had a plenty placed at my disposal, but I do not wish to be under too much obligation in that way. Mr. Hodgkiss gave me 10 dollars and told me to say how much I wanted. I refused any more. Mr. Galpin made me a present of 50 dollars, and Mr. Dawson of Fort Benton [loaned] 100 dollars for which I gave him a note "payable when I am able." I never asked for a cent. They knew my circumstances, however. So you can see that they are friends to me. I also earned 25 dollars at the Treaty. This will keep me from actual want for a time.

I shall ever remain most faithfully yours,

F. V. Hayden[35]

Hayden reached St. Louis late on January 11, 1856. The trip from the mouth of the Judith to St. Louis had taken them, in all, more than two and a half months. Thus ended Hayden's second western tour. Almost twenty months had lapsed since he and Vaughan had left St. Louis aboard the *Genoa* on June 1, 1854. Return to St. Louis marked not only the safe conclusion of a long and perilous journey but also, though he was not yet assured of the fact, the end of Hayden's hand-to-mouth existence in the West. True, he would continue for several more years to pursue what was essentially a "one man natural history survey" in the territories, but henceforth he would not have to operate quite so much on a bare subsistence basis or under circumstances so largely beyond his control. Better times were in the offing.

Epilogue

Thus Ferdinand Hayden's arduous downriver journey of almost three months finally brought him to St. Louis in mid-January 1856. Once there, he was anxious to continue on east to Washington. But first the natural history specimens had to be sorted, culled, and repacked, then sent ahead to Spencer Baird at the Smithsonian. And Hayden discovered with dismay that much of that material was in sorry condition — many specimens so spoiled as to be worthless.

Parts of Hayden's collection still remained in storage at the upper forts, scattered all the way from Fort Pierre to Fort Benton. Nonetheless, the fossils that had reached St. Louis weighed at least six tons. Now they had to be unpacked, roughly classified, and many of them had to be placed in storage pending a final division between Vaughan, Chouteau, and Hayden.

Vertebrate fossils could be sent collect to Spencer Baird in Washington or to Joseph Leidy in Philadelphia, but any shipments of invertebrate and plant fossils might have to be paid for in St. Louis. Hayden was anxious to get his share of the invertebrate and plant fossils laid before the eyes of his friend F. B. Meek, and as a start he paid to send one box of choice invertebrate fossils to Meek at Albany. He also packed other boxes with specimens that he wanted to work on in Meek's company and sent those off to him collect. The remaining fossils — including heavy slabs of fossil plants — had to be left in St. Louis. So a room was prepared at Pierre Chouteau's home where they would not only be stored but would be displayed to the public.

Back when he had been writing from the field, Hayden had seemed almost extravagant in describing some of his finds. Now it turned out that he had been correct. As fellow scientists began viewing Hayden's collection they were elated. Baird found in his shipment the first rodent fossil from the Dakota Badlands and he asked permission to study it (which complicated matters, since Hayden had promised Leidy the description rights for all the vertebrate fossils). Benjamin Shumard wished to investigate the fresh-water invertebrates (which Hayden had intended for Meek). In a shipment of "Saurian material," Joseph Leidy found much that was new and important. George Engelmann said it would take Meek and Hayden years to "work up" just the invertebrates that were new to science.

Fortunately for Hayden, Engelmann was home that winter in St. Louis and-he joined Hayden during the evenings to examine the "Flora Nebraskana." Engelmann's expertise identified many doubtful species and progress was made on a catalog of Upper Missouri plants. Hayden was also busy corresponding with Meek and Hall to plan how they would study the invertebrate fossils. Hall eventually made his facilities available for Meek and Hayden to do that work in Albany. And all the while, Hayden was exploring possible ways to return to the Dakota country for the following summer of 1856.

After a month in St. Louis, Hayden left in mid-February for Washington. There, through the mediation of Baird, Hayden soon received a letter[1] from the Army lieutenant Gouverneur Kemble Warren offering him a position as assistant on an expedition into the "Sioux country" that coming summer. His salary would be $1,000 a year. Hayden accepted the offer immediately and his employment began retroactively as of January 1, 1856. Warren specified that "whatever collections you make during this time are to be for the government of the U.S.," but he added that "if you wish to present a report illustrative of our labors on the geology of the portions of Nebraska Ter. explored by you, I will gladly transmit it to the Department with my report and allow it to take the same chance of publication by Congress. I will also aid you to a limited extent in transporting your collections to such points for examination and comparison as may be proper. I have not money now to authorize any illustrations, but should you gain the assistance of an [*sic*] person through his professional interest to make such, I will endeavor, should my future means justify, to have them recompensed, the result of such labor then becoming the property of the U.S."[1] Warren wanted Hayden to write "a preliminary sketch of the country to be done as soon as possible," and this Hayden did immediately in Washington. That compressed his work with Meek at Albany to a mere five weeks, since he had to be back in St. Louis by April 10th to board the steamboat upriver. Nevertheless, Hayden also managed to visit Leidy in Philadelphia and his own mother in Rochester, New York.

Hayden's appointment by Lt. G. K. Warren was precisely the kind of opportunity he had sought. It ended his life of dire poverty. The appointment also gave him professional standing, a chance to write for publication, and a way to return to "the old stampede ground" (as he called it in a letter to Meek) with guarantees of logistical support.

Hayden thereby ceased to be a neophyte. His collaboration with G. K. Warren in summer 1856 would prove a crucial turning point in his career. No longer would he embark up the Missouri River under makeshift auspices. His foot was now in the federal government's door and that is the opening he would eventually parley into the grand surveys carried out after the Civil War — surveys not only of the Upper Missouri but of the mountains and mining districts of Colorado and Wyoming.

During the rest of 1856 Hayden, in company with Lt. G.K. Warren, returned to the Yellowstone River. In 1857 (again with Warren) he worked around Fort Laramie, the Black Hills, and the Badlands. The year 1858 saw him less in the field, but he did spend much of that fall with F. B. Meek working in Kansas. The years 1859 and 1860 brought another extended government-sponsored journey, this time with Captain William F. Raynold. Spring and summer 1859 took him back to the Yellowstone River and its tributaries. He wintered on the Platte River and the 1860 season saw him bisect what is now western Wyoming, traverse Jackson's Hole and cross into what is now Idaho, descend the Madison River to Three Forks, and also explore other branches of the Upper Missouri River.

Finally in 1862, with the Civil War underway, Hayden's long-planned book was published under the title *Geology and Natural History of the Upper Missouri.* He then enlisted in the Union army and rose to be chief medical officer of Gen. Phil Sheridan's Army of the Shenandoah, leaving the army in May 1865 with the rank of lieutenant colonel by brevet.

The summer and fall of 1866 saw Hayden back in the Badlands collecting again. Starting in 1867 he headed the Geological Survey of Nebraska (which gained statehood that year) and in 1869 that project evolved into the Geological Survey of the Territories, which in turn led ten years later to the U.S. Geological Survey.

The frustrations that Hayden had faced during his earliest forays west during the years 1853, 1854, 1855 — and the countless expedients that he devised during those years — had prepared him well for the achievements to come.

Appendix 1: Ferdinand V. Hayden Timeline, 1845-1856

1845: Hayden entered the Preparatory Department of Oberlin College, Oberlin, Ohio.

1850: Graduated from Oberlin College.

1851-early 1853: In Cleveland studying at the Medical Department of Western Reserve College.

May 9, 1853: Arrived in St. Louis to make preparations for his first Badlands expedition.

May 21, 1853: Departed St. Louis with Meek on the steamboat *Robert Campbell*, bound for Fort Pierre.

June 19, 1853: Arrived at Fort Pierre.

June 21, 1853: Left Fort Pierre to collect fossils in the Badlands.

July 18, 1853: Returned to Fort Pierre.

July 27, 1853: Started downriver in Mackinaw boats with fossil collection.

August 21, 1853: Arrived at St. Louis.

August 23, 1853: Left St. Louis by steamboat.

September 3, 1853: Arrived at Albany, New York.

September 1853: Began studies at Albany Medical College.

December 1853: Finished "course of lectures" at Albany Medical College.

January 1854: Moved into Hall's "office" in Albany, working with Meek.

February 1854: Moved to Rochester, New York.

Early May 1854: Met with Vaughan briefly in New York City; returned to Rochester.

Early May 1854: Made brief trip to Philadelphia; returned to Rochester.

May 13, 1854: Left for St. Louis via Cleveland and Chicago.

May 17, 1854: Arrived in St. Louis.

June 1, 1854: Left St. Louis with Vaughan on the steamboat *Genoa*, bound via Fort Pierre for Fort Union.

July 3, 1854: Arrived at Fort Union.

July 18, 1854: Began the trip up the Yellowstone River bound for Fort Sarpy.

August 15, 1854: Arrived at Fort Sarpy.

September 18, 1854: Left Fort Sarpy heading downriver.

October 6, 1854: Arrived at Council Bluffs, Iowa.

November 26, 1854: Left on horseback for Fort Pierre with stopovers at Kanesville and Sergeant's Bluff.

Mid-December 1854: Arrived back at Fort Pierre.

Late January 1855: Fourteen-day trek with Galpin to Moreau Trading Post and back to Fort Pierre.

February 11, 1855: Began winter trip to the Badlands and Black Hills.

March 22, 1855: Returned to Fort Pierre.

May 7, 1855: Began spring trip to the Badlands.

June 6, 1855: Returned to Fort Pierre from the Badlands.

June 26, 1855: Left Fort Pierre with Vaughan on the *St. Mary*, bound via Fort Union for Fort Benton.

July 11, 1855: Arrived at Fort Union.

July 17, 1855: Left Fort Union to continue to Fort Benton.

August 17, 1855: Arrived at Fort Benton.

October 16-17, 1855: Present at the signing of the "Blackfoot Treaty" near the mouth of the Judith River about 100 miles below Fort Benton.

October 22, 1855: Left the mouth of the Judith River bound for St. Louis via boats to Fort Pierre and then overland via horseback, then carriage, and finally railroad.

January 11, 1856: Arrived at St. Louis.

Appendix 2: The Geologic Time Scale

Era	Period	Age*	Epoch	Dominant life forms
	Quaternary		Holocene	Modern animals
			Pleistocene	Most recent; The "Ice Age"
		2		
				The "Age of Mammals"
			Pliocene	More recent
Cenozoic	Tertiary		Miocene	Less recent
			Oligocene	Slightly recent
			Eocene	Dawn of the recent
			Paleocene	First dawn of the recent
		66		
				The "Age of Reptiles"
	Cretaceous			Wide-spread shallow seas teaming with life
		144		
Mesozoic	Jurassic			Dinosaurs
		208		
	Triassic			Rise of a great variety of reptiles
		245		
	Permian			Mammal-like reptiles
		286		
	Carboniferous			
	Pennsylvanian			Swamp forests forming beds of coal
	Mississippian			Marine invertebrates; crinoids
		360		
Paleozoic	Devonian			Forests of primitive plants
		408		
	Silurian			Marine invertebrates; coral reefs
		438		
	Ordovician			Marine invertebrates; brachiopods, bryozoans
		505		
	Cambrian			Marine invertebrates; trilobites
		570		
Precambrian				
		4000		

*Approximate, in millions of years ago

Notes

Editors' Preface

1. See, for example, these papers by Fryxell: "Thomas Moran's Journey to the Tetons in 1879," *Augustana Historical Society Publications*, No. 2 (1932), pp. 37-46; "Thomas Moran: An Appreciation," *Sierra Club Bulletin* (June 1934), pp. 56-57; "Thomas Moran's Journey to the Mount of the Holy Cross in 1874," *Trail and Timberline*, Colorado Mountain Club, No. 203 (1935), pp. 103-105; "The Thomas Moran Art Collection of the National Parks," *Yosemite Nature Notes*, Vol. 15 (1936), pp. 57-60; "Thomas Moran: Explorer in Search of Beauty," East Hampton, Connecticut, East Hampton Free Library, 1958 [84 pages].

2. Fritiof M. Fryxell, "William H. Jackson: Photographer, Artist, Explorer," *American Annual of Photography* (1939), pp. 1-13.

3. Fritiof M. Fryxell, "Albert Charles Peale (1849-1914): Pioneer Geologist of the Hayden Survey," *Annals of Wyoming*, Vol. 34 (1962), pp. 175-192.

4. Noel Evans and Fritiof M. Fryxell, "Jesse V. Howell (1891-1971)," *American Association of Petroleum Geologists Bulletin*, Vol. 55 (1971), pp. 1885-1887; F. M. Fryxell, "Memorial to Jesse V. Howell, 1891-1971," *Geological Society of America Memorials*, Vol. III (1974), pp. 109-115.

5. Memo from Fryxell to Howell, February 17, 1964; in the Fryxell Papers, Augustana College Library, Special Collections (hereafter cited as ACL), Box 32, Folder 54.

6. Mike Foster, *Strange Genius: The Life of Ferdinand Vandeveer Hayden* (Niwot, Colorado: Roberts Rinehart Publisher, 1994); James G. Cassidy, *Ferdinand V. Hayden: Entrepreneur of Science* (Lincoln: University of Nebraska Press, 2000).

7. Mike Foster, "Ferdinand Vandeveer Hayden as a Naturalist," *American Zoologist*, Vol. 26, pp. 343-349.

8. Grant proposal to the National Science Foundation, 1959; Fryxell Papers, ACL; Box 34, Folder 4.

9. Fryxell Papers, ACL; Box 35.

Chapter 1: The Boy of the Class

1. Kirkland to Baird, January 10, 1853; Spencer Baird Papers, Smithsonian Institution Archives (hereafter SIA), RU 7002; transcript copy also in the Fryxell Papers, ACL, Box 29, Folder 36.

2. Hayden to Baird, February 6, 1853; SIA, RU 7002. This letter is number 351 in Baird's personal file at SIA, which contains over four hundred letters from Hayden to Baird. Baird had joined the Smithsonian staff only two years earlier, in the fifth year after its founding (1846). The last sentence of the first paragraph of this letter has been quoted by both of Hayden's two recent biographers: see Mike Foster, *Strange Genius*, p. 39; and James G. Cassidy, *Ferdinand V. Hayden: Entrepreneur of Science*, p. 41.

3. Oesher, *The Smithsonian Institution*, pp. 89-91; Herber, "Spencer F. Baird: A World Famous Naturalist from Cumberland County."

4. Hayden's paternal lineage was investigated by Mrs. I. W. Albrecht of the Edwin Smith Historical Library, Westfield, Massachusetts. Correspondence between Mrs. Albrecht and J. V. Howell is in the Hayden Collection at the American Heritage Center, University of Wyoming (hereafter UW), Accession 1684; photocopies are available at ACL, Box 30, Folder 6.

5. *Rochester* [N.Y.] *Republican*, August 30, 1831, p. 2, column 6.

6. Based on an account by William B. Thom that was filed with the Colorado Historical Society and was published in *The Eagle Valley Enterprise*, Eagle, Colorado, issues for December 30, 1921 and January 6, 1922.

7. Charles A. White, "Memoir of Ferdinand Vandiveer Hayden, 1839 [*sic*]-1887," National Academy of Science Biographical Memoirs 3 (1895), 397-398. Many years later (1888), mindful of the difficulties alluded to by C. A. White and of Hayden's own gratitude to his Alma Mater, Mrs. Emma W. Hayden presented the main part of her late husband's library to Oberlin College, along with an endowment of $1,000, the income from which, she specified, should be "available for the aid of self-supporting young men." In her will she named Oberlin College the residuary legatee of her estate, in consequence of which the College, in 1936, accepted Mrs. Hayden's bequest of $16,000, the income from which was to be used for awards "to be known as the Ferdinand V. Hayden Fellowships or Scholarships." This money was placed in the scholarship fund of the College, and the income appears to have been used for the purpose specified, but no Hayden Fellowship or Scholarship seems ever to have been set up; at least none is indicated in the catalog of Oberlin College.

8. Rev. Thomas H. Robinson, "Reminiscences of the Class of 1850" (Reunion of 1870); Oberlin College Archives. This and other insights into Hayden's Oberlin experience were submitted to Dr. Albert A. Wright in 1888, shortly after Hayden's death, by some of his surviving classmates. They are preserved in the alumni records of Oberlin College and in SIA, where typed copies were deposited by A. C. Peale.

9. Hayden to Allen, May 11, 1870; Oberlin College Archives.

10. Hayden to Allen, September 2, 1871; Oberlin College Archives.

Chapter 2: Recruit for Science

1. John J. Stevenson, "John Strong Newberry," *American Geologist*, Vol. 12 (1893), pp. 1-25.

2. Ibid., p. 2.

3. John M. Clarke, *James Hall of Albany, Geologist and Paleontologist*, p. 95.

4. Stevenson, p. 3. Adolphe Theodor Brongniart (1801-1876) was a French paleobotanist, best known for his description of Carboniferous flora.

5. Ibid., p. 3.

6. Clarke, p. 244. See also E. D. Cope, "F. V. Hayden," pp. 110-113.

7. Newberry to Hayden, September 21, 1853; Western Reserve Historical Society, Cleveland, Ohio, ms. 3154.

8. Herbert Thoms, *The Doctors Jared of Connecticut.*

9. Clarke, pp. 236-237.

10. George P. Merrill, *The First Hundred Years of American Geology*, pp. 527-528; also Ellis Yochelson, "Fielding Bradford Meek," pp. 255-256; and Nelson and Fryxell (1979).

11. David Dale Owen (1807-1860) lived in New Harmony, Indiana, where a utopian experiment had been established by his father Robert Owen. His most enduring work was a federally funded survey of Wisconsin, Iowa, and Minnesota. He also directed geological surveys of Indiana, Kentucky, and Arkansas. (See Hendrickson, 1943; and Rabbitt, 1979.)

12. Hall to Meek, December 15, 1851; SIA, RU 7062; a transcript is also located in the Fryxell Papers, ACL, Box 29, Folder 32.

13. Clarke, pp. 243-244; also see Finley, "Acrimony in Albany."

14. For a biographical sketch of Evans, see Erwin F. Lange, "Dr. John Evans (1812-1861), U.S. Geologist to the Oregon and Washington Territories."

15. Clarke, pp. 245-246.

16. For a biographical sketch of Alexander Culbertson, see Anne McDonnell, ed., "The Fort Sarpy Journal," pp. 240-242.

17. See John F. McDermott (ed.), in Thaddeus A. Culbertson, "Journal of an Expedition to the Mauvaises Terres and the Upper Missouri in 1850" [1952], p. 1. Also see Cleophas C. O'Harra, "The White River Badlands," p. 161.

18. See Hiram A. Prout, "A Description of a Fossil Maxillary Bone of Paleotherium from Near White River." Also see O'Harra, pp. 23, 161.

19. McDermott (ed.), in Culbertson [1952], p. 1.

20. Joseph Leidy, *The Ancient Fauna of Nebraska.*

21. David Dale Owen, *Report of a Geological Survey of Wisconsin, Iowa, and Minnesota*, pp. 194-206.

22. See McDermott (ed.), in Culbertson [1952], pp. 2-3; also McDonnell, ed., p. 241.

23. Thaddeus A. Culbertson, "Journal of an Expedition to the Mauvaises Terres and the Upper Missouri in 1850" [1851], Smithsonian Institution, Annual Report 5, pp. 84-132.

24. Culbertson [1851].

25. Merrill, *The First Hundred Years of American Geology*, p. 273.

26. The Cretaceous is the last time period of the Mesozoic Era, the Age of Reptiles (see Appendix 2). At this time a shallow sea covered the middle of North America from the Gulf of Mexico to the Arctic Ocean. In the Upper Missouri Country the sediments deposited in this sea consist largely of dark colored shale (compacted mud) hundreds of feet thick and containing fossil oysters and many other varieties of marine life.

27. Owen, pp. 535-572.

28. O'Harra, p. 24.

29. Merrill, *The First Hundred Years of American Geology*, p. 276.

30. Owen, pp. 196-197.

Chapter 3: Hayden is Hired

1. Charles Whittlesley (1808-1886) conducted geological studies in the Lake Superior region, Wisconsin, Pennsylvania, and Ohio. See the *American Geologist*, Vol. 4 (1889), pp. 258-268.

2. Hayden to Hall, January 29, 1853; James Hall correspondence; New York State Archives (hereafter NYSA), Accession No. KW 12835.

3. Hall to Engelmann, February 5, 1853; George Engelmann Papers, Missouri Botanical Garden Archives, St. Louis, Missouri (hereafter MBGA); typescript copy at ACL, Box 29, Folder 29.

4. Pierre Chouteau (1789-1865) was an astute businessman, son of a prominent St. Louis family; he was successful in the fur trade, and expanded his business interests and spent much of his career in New York City. "Chouteau was liberal toward scientific expeditions seeking to go up the country, and he contributed in no small degree to their success" (Chittenden, *The American Fur Trade of the Far West*, Vol. 1, pp. 382-384).

5. H. A. Prout (1808-1862) was a successful St. Louis physician.

6. Dr. Henry King surveyed the geology along the Osage River in Missouri, the only completed study of a more extensive Missouri geological survey that was authorized in 1833. Dr. King also served as president of the Western Academy of Natural Sciences.

7. Engelmann to Hall, February 25, 1853; NYSA; typescript copy at ACL, Box 29, Folder 25.

8. Jacob Whitman Bailey (1811-1857) was professor of chemistry, mineralogy, and geology at West Point. See Elliott, p. 21.

9. Bailey to Hall, March 7, 1853; NYSA; typescript copy at ACL, Box 29, Folder 29.

10. Albert C. Koch (1804-1867) was a fossil collector and showman. He emigrated from Germany in 1826 and established a museum in St. Louis. He was the first to collect and mount (incorrectly) a mastodon. Eventually it was reconstructed and put on display in the British Museum. See *Geotimes*, Vol. 44 (March 1999), pp. 20-24 for an account of Koch's misadventures.

11. Hall to Engelmann, March 27, 1853; MBGA; typescript copy at ACL, Box 29, Folder 29.

12. Newberry to Hall, March 2, 1853; NYSA; typescript copy at ACL, Box 29, Folder 29.

13. Mike Foster, *Strange Genius*, p. 44, speculates that this unidentified friend is Dr. H. A. Prout of St. Louis. J. V. Howell suggested that it might be Meek.

14. Newberry to Hall, March 23, 1853; NYSA; typescript copy at ACL, Box 29, Folder 29.

15. Hayden to Hall, March 3, 1853; NYSA; Merrill Papers, Library of Congress.

16. Hayden to Hall, March 19, 1853; NYSA; Merrill Papers, Library of Congress.

17. Hayden to Hall, March 30, 1853; NYSA.

Chapter 4: Launching an Expedition

1. Engelmann to Hall, March 19, 1853; NYSA; typescript copy at ACL, Box 29, Folder 29.

2. George Clinton Swallow (1817-1899); his principal geological contributions resulted from his work as state geologist of Missouri, and later of Kansas. See the *American Geologist*, Vol. 24 (1899), pp. 1-6.

3. Engelmann to Hall, March 27, 1853; NYSA; typescript copy at ACL, Box 29, Folder 29.

4. Hall to Engelmann, April 10, 1853; MBGA; typescript copy at ACL, Box 29, Folder 29.

5. Prout to Hall, April 27, 1853; NYSA; typescript copy at ACL.

6. Abram Litton (1814-1901) was an assistant to G. C. Swallow on the geological survey of Missouri. See Merrill [1920], p. 704.

7. Hall to Hayden, April 19, 1853; National Archives (hereafter NA), Record Group 57 (hereafter RG 57); typescript copy at ACL, Box 29, Folder 30.

8. Hall to Chouteau, April 20, 1853; NA, RG 57; typescript copy at ACL, Box 29, Folder 28.

9. Hall to Engelmann, April 20, 1853; MBGA; typescript copy at ACL, Box 29, Folder 29.

10. Meek to Hayden, undated; NA, RG 57.

11. William Edmonds Horner (1793-1853) was professor of anatomy and Dean of the Medical School at the University of Pennsylvania.

12. Joseph Henry (1797-1878) made major contributions to science as head of the Smithsonian Institution, promoting scientific investigation; he was also a pioneer in the study of electromagnetism. See Thomas Coulson, *Joseph Henry: His Life and Work.*

13. Leidy to Hall, April 20, 1853; NYSA.

14. Hall to Engelmann, April 23, 1853; MBGA; typescript copy at ACL, Box 29, Folder 29.

15. James Dwight Dana (1813-1895) was professor of geology and mineralogy at Yale University. His interest in crustaceans grew out of his experience with the Wilkes Expedition of 1838-1842. See Daniel C. Gilman, *The Life of James Dwight Dana.*

16. Hall to Hayden, April 24, 1853; quoted from Merrill, *The First Hundred Years of American Geology*, p. 709; typescript copy at ACL Box 29, Folder 30.

17. Hayden to Hall, April 26, April 28, April 29, 1853; NYSA; and Hayden Collection, American Heritage Center, University of Wyoming (hereafter UW), Accession 1684.

18. Hayden to Meek, April 29, 1853; SIA; typescript copy at UW.

19. Hall to Hayden, May 9, 1853; NA; quoted from Merrill, *The First Hundred Years of American Geology*; photocopy at ACL, Box 29, Folder 30.

20. Father Pierre Jean De Smet, S.J. (1801-1873) came to St. Louis from his native Belgium in 1821. He served as a missionary to Indians, first in Council Bluffs and later throughout the Rocky Mountains. Many of his letters to friends and family back home contain vivid descriptions of the American West. See Chittenden [1954], p. 648.

21. Robert Campbell (1804-1879) was a fur trader, a St. Louis businessman, and an Indian Commissioner. See Chittenden [1954], p. 260.

22. John B. Sarpy (1798-1857) was second only to Pierre Chouteau in the leadership of the American Fur Company. See Chittenden [1954], p. 390.

23. Hayden to Hall, May 10, 1853; NYSA; typescript copy at UW.

24. Hall to Chouteau, May 7, 1853; SIA; typescript copy at ACL, Box 29, Folder 30.

25. Major General Isaac Ingalls Stevens (1817-1862) graduated from West Point in 1839 and served in the Mexican War. Appointed governor of Washington Territory in 1853, he was killed during the Battle of Chantilly, September 1, 1862. See Hazard Stevens, *The Life of Isaac Ingalls Stevens.*

26. Hall to Meek, May 11, 1853; SIA, RU 7062; typescript at ACL, Box 29, Folder 32.

27. Joseph Granville Norwood (1807-1895) worked with David Dale Owen on the geological survey of Wisconsin, Iowa, and Minnesota, and served as state geologist in Illinois, before joining the geological survey of Missouri. See *The American Geologist*, Vol. 16 (1895), pp. 69-74; Merrill [1920], p. 707.

28. Benjamin Franklin Shumard (1820-1869) was a physician and paleontologist who served with the Owen Survey and with state surveys in Kentucky, Missouri, and Texas. See the *American Geologist*, Vol. 4 (1889), pp. 1-6.

29. Hayden to Hall, May 16, 1853; NYSA; quoted from Merrill, *The First Hundred Years of American Geology*, p. 696.

Chapter 5: Underway at Last

1. Jean Louis Rudolphe Agassiz (1807-1873); professor of natural history, Lawrence Scientific School, Harvard University, an early proponent of the glacial origin of the "drift." See Edward Lurie, *Louis Agassiz: A Life in Science.*

2. Clarke, *James Hall of Albany, Geologist and Paleontologist*, pp. 246-247.

3. Meek to Hall, May 19, 1853; NYSA; quoted from Merrill, *The First Hundred Years of American Geology*, pp. 697-698.

4. Engelmann to Hall, May 20, 1853; NYSA; typescript at ACL, Box 29, Folder 25.

5. Engelmann to Baird, June 28, 1853; SIA, RU 7002; typescript at ACL, Box 29, Folder 24.

6. Hall to Engelmann, June 15, 1853; MBGA; typescript at ACL, Box 29, Folder 29.

7. Adam August Krantz (1808-1872) was a world-wide dealer in minerals, fossils, and rocks as teaching aids and for museum display. He originally operated from Freiberg, Germany, with subsequent moves to Berlin and Bonn. The company is still in business.

8. Hall to Engelmann, undated; MBGA; typescript at ACL Box 29, Folder 29.

9. Hall to Henry, July 23, 1853; NYSA; typescript at ACL, Box 29, Folder 31.

10. Clarke, p. 249.

Chapter 6: The Trip to Fort Pierre

1. See Henry A. Boller, *Among the Indians: Four Years on the Upper Missouri*; Hiram Martin Chittenden, *History of Early Steamboat Navigation on the Missouri River*; Joseph Mills Hanson, *The Conquest of the Missouri*; Rudolph F. Kurz, "Journal of Rudolph Friederich Kurz"; and Thaddeus A. Culbertson, "Journal of an Expedition to the Mauvaises Terres and the Upper Missouri in 1850" (1850, 1952).

2. Meek's papers are preserved in SIA, RU 7062.

3. Merrill, *The First Hundred Years of American Geology*, pp. 699-709.

4. This limestone is probably Lower Carboniferous, Mississippian, in age; rocks in the lower part of these bluffs may be as old as Ordovician (refer to the Geologic Time Scale in Appendix 2). Progressively younger rock is encountered as one journeys up the Missouri River.

5. The drift deposits consist of unconsolidated clay, silt, sand, and gravel deposited by continental glaciers during the Pleistocene Epoch (refer to the Geologic Time Scale in Appendix 2). In general, the Missouri River marks the southern and western limits of the vast ice sheet that had its center in east-central Canada and covered much of the north-central United States.

6. The term "coal measures" refers to the coal-bearing rocks of Pennsylvanian age (Upper Carboniferous).

7. This is probably the Dakota Sandstone of Late Cretaceous age, not the Potsdam Sandstone, which is Cambrian in age. These sandstones have a very similar appearance, and their relative age can be determined only by careful examination of their contained fossils. See Owen, pp. 187-193.

8. Loess is glacial flour, silt and very fine sand, produced by the grinding action of glacial ice. Some of this sediment is deposited on the floodplains of rivers by water derived from the melting edge of a glacier, and then, during dry seasons, it is blown by the wind across the countryside. On the leeward, eastern, side of the broad Missouri River floodplain in western Iowa the loess is more than 100 feet thick and has been eroded into an unusual landscape, a labyrinth of short, steep hills and valleys.

9. Banvard's Panorama was a scroll-like mural tens of feet in length showing the bluffs bordering the valley as viewed from the river. When simultaneously unrolled at one end and rolled up at the other it provided a vicarious trip down the river. (For a biographical sketch of Banvard, see Palmquist and Kailbourn, pp. 86-88.)

Chapter 7: Fossil Hunting in the Badlands

1. This is the Pierre Shale, compacted mud deposited in the Cretaceous seaway that extended from the Gulf of Mexico to the Arctic Ocean. It lies on the Dakota Sandstone and occurs throughout the northern plains.

2. So-called "desert varnish," a coating of iron and manganese oxides on the surface of porous rocks.

3. Gneiss is a metamorphic rock formed under conditions of high heat and pressure within the earth. Uplift of the earth's crust and erosional removal of overlying rock has brought it to the earth's surface over large areas of Minnesota and Ontario. This boulder was carried here by the continental glacier. It is a "glacial erratic."

4. This is actually the Bad River, a tributary that joins the Missouri River at Fort Pierre. The Little Missouri River heads in northeastern Wyoming and flows north through Theodore Roosevelt National Park and joins the Missouri River in west-central North Dakota.

5. The Niobrara River.

6. Typical invertebrate fossils from the Pierre Shale.

7. Paleotherium beds are the lowest (and oldest) strata of Tertiary age in the Badlands.

8. Selenite is the clear, crystalline form of gypsum ($CaCO3$). The veins are the result of gypsum and chalcedony (finely crystalline quartz) being chemically precipitated from water moving through cracks in the host rock.

9. These reddish layers are now known to be ancient soils.

10. Concretions are nodules of mineral material deposited by water percolating through the rock. They often form around fossils enclosed in the sediment.

Chapter 8: Downiver to St. Louis

1. The memorandum lists the following articles:
   1. Overcoat
   1. Hat
   1. Large Box Medze [Medicine?]
   Half bbl. flour
   Half bbl of pork
   1. Trunk
   1. Box specimens small
   1. Bundle Botanical Specimens
   1. Bunch cactus
   2. Bottles Insects and reptiles etc
   1. Empty box
   Some of the things listed were perhaps left behind in expectation that Meek or Hayden, or both, would presently return to the fort and reclaim them. The specimens may have been left for someone else to take down to St. Louis. Perhaps the bottles of insects and reptiles were left with someone who was willing to collect additional specimens. Doubtless these specimens were to go eventually to Spencer Baird, and the plants were intended for George Engelmann.

2. See Chittenden, *The American Fur Trade of the Far West*, Vol. 1, p. 94.

3. These clays, the Pierre Shale, are very weak and tend to slide down slopes as incoherent blocks.

4. "Glacial erratics."

5. Great Bend is a very large meander loop of the Missouri River located about 70 miles below Fort Pierre. The outcrops of Cretaceous sediments are most likely on the west bank, the southwestern corner of Hyde County, South Dakota.

6. The White River enters the Missouri from the west about 120 miles downstream from Fort Pierre. It is the principal stream draining the Badlands. It is white because it carries a heavy load of white sediment eroded from the Badlands.

7. "Dip" refers to the angle of inclination of the rock layers relative to the horizontal. Sedimentary rock layers are usually deposited horizontal (dip is 00); any departure from horizontal indicates disturbance by movement of the earth's crust, or, as in this case, by landslides.

8. The Bijou Hills are located along the east shore of the Missouri River in Brule County, South Dakota. They rise about 700 feet above the river.

9. Joseph Nicollet (1786-1847) was an astronomer, mathematician, geophysicist, and stratigrapher who emigrated from France in 1832. He prepared maps of portions of Wisconsin, Minnesota, and the Dakotas. See Merrill [1920], p. 316.

10. Scalp Mountain is located about 170 miles downstream from Fort Pierre in Gregory County, South Dakota.

11. A "drift terrace" is a former river floodplain below which the river has eroded a lower, modern, floodplain. It is composed of sediment, mostly gravel and sand, derived from the melting edge of the continental glacier.

12. Towerhill rises above the right bank of the Missouri River in Gregory County, South Dakota.

13. The Niobrara River, a major tributary entering the Missouri from the west.

14. Daurion Hill is located about 10 miles upstream from Yankton, South Dakota.

15. James River, a major tributary entering the Missouri from the north.

16. Vermillion River, another major tributary entering the Missouri from the north.

17. Elk Horn Point is a prominent meander loop of the Missouri River. The town of Elk Point, South Dakota now lies on this extensive prairie.

18. The Big Sioux River joins the Missouri at Sioux City, Iowa, where it forms the boundary between Iowa and South Dakota.

19. At several places in Dixon County, Nebraska, the Missouri River flows along the very foot of its bluffs. Any, or all, of these could be considered to be Dixon's Bluff. These slopes are unstable, subject to slump and landslide, perhaps accounting for the discrepancy between the observations of Nicollet and those of Meek. A portion of this landscape is preserved in Ponca State Park.

20. The gravesite of Sergeant Floyd, the only fatality suffered by the Lewis and Clark expedition.

21. Hall to Meek, August 12, 1853; SIA, RU 7062; typescript copy at ACL, Box 29, Folder 32.

22. Hall to Hayden, August 12, 1853; NA, RG 57; transcript copy at ACL, Box 29, Folder 30.

Chapter 9: Aftermath of the Journey

1. Jules Marcou was a protégé of Louis Agassiz, and his 1853 map, followed by subsequent revisions, contributed circuitously to Meek's disenchantment with James Hall. See Clifford M. Nelson, "Toward a Reliable Geologic Map of the United States," pp. 60-64.

2. Meek to Engelmann, September 17, 1853; MBGA, Book M; typescript copy at ACL, Box 29, Folder 40.

3. Engelmann to Meek, September 30, 1853; SIA, RU 7062; typescript copy at ACL, Box 29, Folder 27.

4. Leidy to Hall, October 1, 1853; NYSA; typescript copy at ACL, Box 29, Folder 37.

5. Hall to Hayden, April 22, 1856; NYSA; typescript copy at ACL, Box 29, Folder 30.

6. Clarke, *James Hall of Albany, Geologist and Paleontologist*, p. 251.

7. Hall and Meek [1856].

8. Merrill, *The First Hundred Years of American Geology*, p. 502.

9. Hall and Meek had entered into a written contract on June 9, 1852, at Meek's own request, evidently because at the time he felt the need for more security in his position. In this contract, Meek agreed to remain at Albany as Hall's assistant for four years (Merrill [1924], p. 723). After the 1853 expedition to the Badlands, G. C. Swallow, in organizing the survey of Missouri, made overtures to Meek to secure his services and urging him to seek a release from his contract with Hall. According to Clarke, Meek was much tempted by Swallow's attractive offer, and begged Hall to release him:

> ... but Meek had now become acquainted with Hall's procedures and collections and to release him would have been grossly unfair to Hall himself. Mr. Hall suggested to Swallow that he take one-half of Meek's time, failing which he would volunteer to pay Mr. Meek as large a sum as Swallow offered, but at any rate the condition of the Paleontology [of New York] would not permit his full release. Mr. Meek did go to Missouri on this part-time arrangement, but it all ended in turmoil. I have put down so much, because in one form or another, Hall's troubles with his assistants became historic. At this time Hall was the personification of superabounding physical vigor capped by a surpassing ambition and a burning irascibility. He actually terrorized those who came in conflict with him personally or entered as competitors upon his field. It is a fact that he administered al fresco pummeling to his men servants and he did not hesitate to make himself impressive on occasion by snatching a shotgun off the rack over his table. Mr. Meek was the witness of these displays and he was frightened, though in many ways his chief was considerate of him and gave him a large opportunity for increasing his knowledge...; yet Meek, once away safely in Missouri, refused to come back to the terrors of this despot in science. That Meek was thoroughly scared is unquestionable; his fear became greater when Hall, scandalized and outraged by the literal violation of the agreement between them, dipped his pen in vitriol and denounced, to such men as Joseph Henry, his assistant in most unmeasured terms. It was an angry and distressing episode and it brought upon Hall only reproaches and warnings from his advisors and equals; but it was his way, and to him the only way when he believed an injustice was done him. For years he nursed his wrath and the expostulations of those who dared to expostulate taught him nothing. The flame died down in time, but it was sure to flare again when the spark was struck. Such acerbity was all a part of the man in these burning years, and it must be estimated in the sum of his doings" (pp. 252-253).

To Clarke's account the following may be added: The experience that Meek gained in the summer of 1853, as well as the offers made by Evans and Swallow, undoubtedly increased his sense of self-reliance, and contributed to his determination to secure an early termination of his contract with Hall, and so become free to make his own way in paleontology. While negotiating with Swallow, he also wrote to Spencer Baird, "I would fall under many obligations to you, if you would use your influence to procure for me an appointment in some of the government Exploring expeditions." He expressed confidence that Hall would "have the magnanimity" to release him in the event that he should receive "a much better offer" (Meek to Baird, September 18, 1853; SIA, RU 7002; typescript copy at ACL, Box. 29, Folder 39). Baird replied: "There is nothing on foot now as far as I know, but should the Institution be called upon for such a purpose, I will with much pleasure keep your name before Prof. Henry as a candidate. We are well aware of your availability from the reports of Dr. Evans and others, and are well assured that you would admirably fill the position you desire. You would of course be bound by your engagement with Prof. Hall, until its expiration or a release by him, but I think with you Prof. Hall would act magnanimously in the event of a better offer being made you. He has always acted with fairness and liberality towards those connected with him, as far as my knowledge extends" (Baird to Meek, October 3, 1853; SIA, RU 7062; typescript copy at ACL, Box 29, Folder 39). By late 1854 Meek had fairly well extricated himself from Hall's grasp, and Hall sourly told George Englemann that "My Mauvaises Terres expedition has proved a very expensive affair, and has been the means also of taking Mr. Meek from me, who is now on the Missouri Survey" (Hall to Englemann. December 12, 1854; MBGA; typescript copy at UW.)

10. Hall to Hayden, April 22, 1856; NYSA; typescript copy at ACL, Box 29, Folder 30.

11. "You returned and by my advice [took] the early course of lectures in Albany, instead of giving two months to me, as we had before agreed upon and which was still due me. I wished to facilitate your progress, and was quite willing to do all in my power." (Hall to Hayden, April 22, 1856; NYSA; typescript copy at ACL, Box 29, Folder 30.)

12. Newberry to Hayden, September 21, 1853; NA, RG 57.

13. Albany Medical College, Catalog and Circular, Fall Term 1853.

14. William Frederick Norwood, *Medical Education in the United States before the Civil War*, pp. 159-163.

15. Clarke, pp. 30, 191, 237.

16. Norwood, p. 163.

17. Albany Medical College, Catalog and Circular, Fall Term 1853.

18. Ibid.

19. The subject of Hayden's thesis has not been found in the records of Albany Medical College.

20. Hayden to Baird, January 8, 1854; SIA, RU 7002; typescript copy at UW.

21. Meek's Diary.

22. Merrill, *The First Hundred Years of American Geology*, p. 524.

Chapter 10: Call of the Wilderness

1. Lesley to Rutherford B. Hayes, February 19, 1879; J. Peter Lesley Papers, American Philosophical Society.

2. Leidy to Hall, December 19, 1853; NYSA; typescript copy at ACL, Box 29, Folder 37.

3. Hayden to Baird, January 8, 1854; SIA, RU 7002; typescript copy at UW.

4. Hayden to Leidy, January 11, 1854; Academy of Natural Sciences of Philadelphia, Manuscript Archives, Collection 1-B (hereafter ANSP).

5. Leidy to Hayden, February 28, 1854; NA, RG 57.

6. George Catlin (1796-1872), an artist who specialized in portraits of Indians of the American West. He was born in Wilkes-Barre, Pennsylvania, studied law, but later turned to art. He began his western travels in the early 1830s.

7. Maximilian, Alexander Philipp, Prince of Wied-Neuwied (1792-1867); a student of natural history and ethnology. In the company of artist Karl Bodmer he visited the Upper Missouri Country, and beyond, during a thirteen-month excursion beginning in April of 1833. See Thomas and Ronnefeldt, eds. *People of the First Man* (New York: E.P. Dutton, 1976).

8. Hayden to Baird, January 27, 1854; SIA, RU 7002; typescript copy at UW.

9. Hayden to Baird, February 8, 1854; SIA, RU 7002; typescript copy at UW.

10. Meek to Hayden, March 4, 1854; NA, RG 57; typescript copy at UW.

11. Horace Greeley (1811-1872) was a largely self-educated newspaper editor. He founded the *New York Tribune* in 1841, which brought him national prominence. He was nominated for President by the Democratic Party in 1872 and was soundly defeated by incumbent president Ulysses S. Grant.

12. Hall to Hayden, March 1, 1854; NA, RG 57; typescript copy at ACL, Box 29, Folder 30.

13. Hayden to Meek, March 4, 1854; SIA, RU 7062; typescript copy at UW.

14. Hayden to Baird, March 5, 1854; SIA, RU 7002; typescript copy at UW.

15. Hayden to Baird, March 20, 1854; SIA, RU 7002; typescript copy at UW.

16. Meek to Hayden, March 21, 1854; NA, RG 57; typescript copy at UW.

17. Hayden to Meek, March 27, 1854; SIA, RU 7062; typescript copy at UW.

18. Meek to Hayden, March 29, 1854; NA, RG 57; typescript copy at UW.

19. Hayden to Engelmann, April 5, 1854; MBGA; typescript copy at UW.

20. Hayden to Baird, April 11, 1854; SIA, RU 7002; typescript copy at UW.

21. Swallow to Hayden, April 6, 1854; Iowa Historical Society, Des Moines, H-3244, Hough Collection.

22. The Gadsden Treaty provided for the purchase from Mexico of 45,500 square miles south of the Gila River in Arizona and New Mexico for the sum of $10,000,000. The treaty and sale was negotiated by James Gadsden (1788-1858), whose varied career included graduating from Yale, extended service in the army, railroad promoter, and diplomat. Captain R. B. Marcy had traveled through this area in 1849 on the way to southern California and reported that this would be a relatively easy route for a southern railroad. Sectional rivalries in Congress delayed ratification of the Gadsden Purchase until 1854.

23. Hayden to Baird, April 14, 1854; SIA, RU 7002; typescript copy at UW.

24. Hayden to Baird, April 26, 1854; SIA, RU 7002; typescript copy at UW.

25. Hayden to Baird, May 5, 1854; SIA, RU 7002; typescript copy at UW.

26. This memo is filed with the Baird papers at SIA, RU 7002; typescript copy at UW.

27. This memo likewise is filed with the Baird papers at SIA, RU 7002; typescript copy at UW.

28. Hayden to Baird, May 13, 1854; SIA, RU 7002; typescript copy at UW.

29. Hayden to Leidy, May 13, 1854; ANSP; typescript copy at UW.

30. Leidy to Hayden, May 15, 1854; NA, RG 57; typescript copy at UW.

31. Hayden to Baird, May 19, 1854; SIA, RU 7002; typescript copy at UW.

32. John Torrey (1796-1873), a chemist by training and vocation, was America's leading botanist in the mid-nineteenth century. Torrey was a mentor of Asa Gray (1810-1888), also a botanist and Professor of Natural History at Harvard University.

33. Hayden to Baird, May 25, 1854; SIA, RU 7002; typescript copy at UW.

34. The location of this contract is unknown.

35. Chalcedony is microcrystalline quartz (SiO_2). In the Badlands it occurs as a chemical precipitate filling vertical cracks in the sedimentary layers.

36. Meek to Hayden, May 4, 1854; NA, RG 57; typescript copy at UW.

37. Merrill, *The First Hundred Years of American Geology*, p. 528.

38. Kirkland to Baird, June 27, 1854; SIA, RU 7002; typescript copy at UW.

Chapter 11: To Fort Sarpy on the Yellowstone

1. Alfred J. Vaughan, "Report of Alfred J. Vaughan, 1854," pp. 287-297.

2. Flora Warren Seymour, *Indian Agents of the Old Frontier*, p. 1.

3. Anne McDonnell, ed., "The Fort Sarpy Journal," p. 272.

4. Charles Larpenteur, *Forty Years a Fur Trader on the Upper Missouri*, p. 346.

5. McDonnell, ed., pp. 272-273.

6. John C. Ewers, *The Blackfeet*, pp. 228-229.

7. Vaughan continued as Indian agent to the Upper Missouri tribes for four years, until 1857, when he petitioned for transfer and became agent to the Blackfeet. In this position he established the agency at Sun River, and he held office until 1861, when the Civil War brought about his retirement, as it did that of many government officials who were sympathetic to the South or suspected of being so. John C. Ewers gives an excellent account of Vaughan's constructive work among the Blackfeet (see Ewers, pp. 228-235).

   In 1855 a son of Vaughan's ascended the Missouri River with Colonel Alfred Cumming. Cumming wrote in his report, "Mr. Vaughan, Jr., and Mr. Kennedy accompanied me to the Judith." Evidently this was the son's first and only visit to the Upper Missouri region. Culbertson remarked that "young Vaughan's Virginia blood" was aroused by a hostile act on the part of some Indian.

   In 1858, when the talented young German-American artist Carl Wimar made the first of his three river trips to the Upper Missouri for the purpose of recording the life and customs of the tribes in that region, he set out in company with Vaughan but soon separated from him for reasons stated in a letter published in the *Duesseldorf Journal*: "During the month of May I commenced my tour to the Indian country, accompanied by the United States Indian Agent, Col. Vaughan, but was informed when we arrived at Sioux City that I had chosen rather a dangerous companion, as considerable hostility existed between that officer of the Government and many of the most warlike tribes. Taking this matter well into consideration, and after consulting with my traveling companion, Capt. Konstantin Blandowsky . . . it was considered a better policy to join ourselves to the American Fur Company, who were in better odor among the nations. This we accordingly did, but were not so much benefited by the change as we anticipated, for the agent still pressed himself upon us" (Hodges, *Carl Wimar*, pp. 17-18).

   Vaughan continued to be a prominent figure in the West for a number of years after leaving the Indian Service. In 1862 he and his Indian wife were passengers on the *Shreveport*. Another passenger on this boat, W. C. Gillette, told this incident concerning the couple: "Major Vaughn was one of the passengers. He was formerly an Indian agent under President Buchanan and had with him his Indian wife and child. Her relatives lived in the vicinity of Fort Pierre. It appears that the Major had purchased at St. Joseph for his wife an elegant silk gown, brocaded with satin figures. She went on shore for a visit with her relatives, and with them went on a berrying expedition attired in this gown. When she returned this garment was a sight to behold, and the Major, using language more forcible than polite, declared that hereafter she should be clad only in the regulation Indian blanket" (McDonnell, ed., p. 272).

The noted Belgian Jesuit priest, Father Pierre Jean De Smet, baptized a daughter of Vaughan, Fanny, four months old, on board the *Yellowstone*, July 11, 1864 (McDonnell, ed., p. 272).

In the summer of 1868, Vaughan was back in Montana when the government made treaties with the Blackfeet, Gros Ventres, and Crows. He did special work for the U.S. Commissioner among these tribes, and signed the three treaties as a witness. He died in Marshall County, Mississippi, in June 1871, aged 70 years (McDonnell, ed., p. 273).

8. Hayden to Baird, June 4, 1854; SIA, RU 7002; typescript copy at UW.

9. Hayden to Baird, June 11, 1854; SIA, RU 7002; typescript copy at UW.

10. Alfred J. Vaughan [1854], p. 287.

11. Vaughan [1854], p. 288.

12. The Treaty of Fort Laramie was concluded on September 17, 1851. It bound the various tribes of the upper Missouri country to peace among themselves, delineated the area belonging to each tribe, and recognized the right of the United States to "establish roads, military and other posts, within their respective territories."

13. Vaughan [1854], p. 288.

14. Hayden to Meek, July 1, 1854; SIA, RU 7065; typescript copy at UW.

15. Hayden to Baird, July 2, 1854; SIA, RU 7002; typescript copy at UW.

16. Hayden to Baird, July 3, 1854; SIA, RU 7002; typescript copy at UW.

17. Chittenden, *The American Fur Trade of the Far West*, Vol. 2, p. 959; also Athearn, *Forts of the Upper Missouri*, pp. 19-22.

18. Isaac Ingalls Stevens, "Correspondence of the Office of Explorations and Surveys concerning Isaac Stevens' Survey of the Northern Route for the Pacific Railway," pp. 85-86.

19. Vaughan [1854], pp. 289-292.

20. Chittenden, *The American Fur Trade of the Far West*, Vol. 2, p. 965.

21. Hayden to Meek, October 6, 1854; SIA, RU 7062; typescript copy at UW.

22. Hayden to Meek, November 11, 1854; SIA, RU 7062; typescript copy at UW.

23. Hayden to Baird, November 11, 1854; SIA, RU 7002; typescript copy at UW.

24. Hayden to Meek, March 25, 1855; SIA, RU 7062; typescript copy at UW.

25. McDonnell, ed., pp. 191-192.

26. Vaughan [1854], p. 292.

27. Vaughan [1854], p. 293.

28. Vaughan [1854], pp. 292-293.

29. Vaughan [1854], p. 295.

30. Vaughan [1854], pp. 296-297.

31. Hayden to Meek, March 25, 1855; SIA, RU 7062; typescript copy at UW.

32. Vaughan [1854], pp. 293-295.

33. Meek to Hayden, August 13, 1854; NA, RG 57; typescript copy at UW.

34. Hayden to Meek, October 6, 1854; SIA, RU 7002; typescript copy at UW.

35. Hayden to Meek, November 11, 1854; SIA, RU 7002; typescript copy at UW.

Chapter 12: Visit to the Settlements

1. Hayden to Baird, November 11, 1854; SIA, RU 7002; typescript copy at UW.

2. J. Peter Lesley, "Obituary Notice of Ferdinand Vandiveer Hayden, M.D, Ph. D., L.L.D.," pp. 59-61.

3. Hayden to Baird, November 11, 1854; SIA, RU 7002; typescript copy at UW.

4. Hayden to Meek, November 11, 1854; SIA, RU 7062; typescript copy at UW.

5. Hayden to Engelmann, November 11, 1854; MBGA; typescript copy at UW.

6. Hayden to Leidy, November 11, 1854; ANSP, Collection 1-B; typescript copy at UW.

7. Hayden to Baird, November 17, 1854; SIA, RU 7002; typescript copy at UW.

8. Hayden to Engelmann, November 26, 1854; MBGA; typescript copy at UW.

9. Col. A. R. Gilmore was one of the early settlers in Douglas County, Nebraska; he served as Receiver in the Land Office in Douglas County in the mid-1850s.

10. Senator Stephen A. Douglas, the "Little Giant." Lawyer, statesman, orator; as a Senator, Douglas supported the Missouri Compromise of 1850 and sponsored the Kansas-Nebraska Act which gave the voters of each of those two territories the opportunity to vote about whether to become a slave state or a free state.

11. Hayden to Baird, November 26, 1854; SIA, RU 7002; typescript copy at UW.

12. Ibid.

13. General George Wallace Jones (1804-1896) organized the Iowa Territory in 1838 and joined A. C. Dodge as one of Iowa's first senators.

14. Augustus Caesar Dodge (1812-1883); served with G. W. Jones as one of Iowa's first senators.

15. Hayden to Baird, December 6, 1854; SIA, RU 7002; typescript copy at UW.

16. Hayden to Baird, February 9, 1855; SIA, RU 7002; typescript copy at UW.

Chapter 13: First Winter in the West

1. Hayden to Engelmann, December 31, 1854; MBGA; typescript copy at UW.

2. Ibid.

3. Ibid.

4. Hayden to Baird, February 9, 1855; SIA, RU 7002; typescript copy at UW.

5. The report of Malcolm Clark's death proved unfounded. Clark did become a victim of Indian violence, but not until years later, on August 17, 1869. See Chittenden, *The American Fur Trade of the Far West*, Vol. 1, p. 394; and McDonnell, ed., "The Fort Sarpy Journal," p. 263.

6. Hayden to Baird, February 9, 1855; SIA, RU 7002; typescript copy at UW.

7. Hayden to Leidy, February 9, 1855; ANSP; typescript copy at UW.

8. This contract, referred to by Hayden in several letters, has not been found. However, the substance of it is contained in the letter of April 9, 1855, to Baird, signed by both Hayden and Vaughan.

9. Hayden to Meek, February 10, 1855; SIA, RU 7062; typescript copy at UW.

10. Hayden to Meek, March 25, 1855; SIA, RU 7062; typescript copy at UW.

11. Hayden to Meek, April 5, 1855; SIA, RU 7062; typescript copy at UW.

12. Hayden to Engelmann, April 5, 1855; MBGA; typescript copy at UW.

13. Hayden to Baird, April 6, 1855; SIA, RU 7002; typescript copy at UW.

14. G. K. Warren, "Explorations in the Dacota Country in the Year 1855," pp. 68, 70-71.

15. Vaughan and Hayden to Baird, April 9, 1855; SIA, RU 7002; typescript copy at UW.

16. Hayden to Baird, mid-April, 1855; SIA, RU 7002; typescript copy at UW.

17. This comment is probably in reference to the ceremonies that were to be conducted at the signing of the Blackfeet Treaty to be held in mid-August, 1855.

18. Hayden to Baird, May 10 [June 10], 1855; SIA, RU 7002; typescript copy at UW.

19. Warren [1855], pp. 66-79.

Chapter 14: Spring Tour to the Badlands

1. G. K. Warren, "Explorations in the Dacota Country in the Year 1855," 34th Congress, 1st Session, 1856, Senate Executive Document 76. (Hayden's report is contained in Appendix E, pp. 66-79. This present chapter reprints pages 71-76 of Hayden's report. Two maps accompanied his report.)

2. Use of the word "mud" in connection with the origin of the sand hills (dunes) appears to be a typographical error. Sand hills are most often produced by wind. There are, however, several other possible explanations for the use of the word "mud" in this connection.

Chapter 15: To Fort Benton and Return

1. Charles Christopher Parry (1823-1890), a botanist and long-time resident of Davenport, Iowa, earned an M.D. at Columbia College, New York City in 1846. While there he studied botany with John Torrey, the celebrated botanist. Parry served as botanist on numerous state and territorial surveys.

2. Meek to Hayden, January 6, 1855; NA, RG 57.

3. William S. Harney (1800-1889) served in the U. S. Army from 1818 until retirement in 1863. He advocated peaceful resolution of conflicts with the Indians.

4. F. T. Wilson and Charles Edmund DeLand, *Old Fort Pierre and Its Neighbors*, p. 278; see also Athearn, pp. 23-27.

5. Quoted in Wilson and DeLand, p. 278.

6. Quoted in Wilson and DeLand, p. 279.

7. Wilson and DeLand, p. 282.

8. Frequently called the *St. Mary's*, "a side-wheeled mountain boat, sank in the Missouri River below Nebraska City on September 4, 1858, on a trip to Fort Union" (McDonnell, ed., "The Fort Sarpy Journal," pp. 289-290).

9. Ewers, *The Blackfeet*, pp. 205-225.

10. Vaughan [1855], p. 392.

11. On June 7th according to Goetzmann, *Army Explorations in the American West*, p. 407.

12. Goetzmann, p. 410; also see Warren [1855], pp. 38-39.

13. Vaughan [1855], p. 393. All subsequent quotations regarding the voyage of the *St. Mary* are from this source.

14. Quoted in McDonnell, ed., p. 132.

15. Hayden to Baird, July 10, 1855; SIA, RU 7002; typescript copy at UW.

16. Hayden to Engelmann, July 11, 1855; MBGA; typescript copy at UW.

17. Hayden to Baird, July 12, 1855; SIA, RU 7002; typescript copy at UW.

18. Hayden to Meek, July 17, 1855; SIA, RU 7062; typescript copy at UW.

19. Quoted in McDonnell, ed., p. 41.

20. Ibid.

21. Ewers, p. 213.

22. Partoll, "Blackfoot Indian Peace Council," p. 199.

23. Hazard Stevens, *The Life of Isaac Ingalls Stevens*, Vol. 2, p. 116.

24. Ewers, pp. 224-225.

25. The detailed account of the Council written by Governor Steven's son Hazard Stevens, who attended the meeting as a boy of 13, is well-written and informative but scarcely unbiased on the subject of the differences between Stevens and Cumming. The youthful Hazard Stevens described Cumming as "a large, portly man, pompous, and full of his own importance," who "attempted to arrogate to himself practically sole and exclusive authority." Hazard Stevens, Vol. 2, p. 102.

26. Partoll, p. 199.

27. Dawson to St. Louis Office, October 16, 1855.

28. Partoll, p. 207.

29. Meek to Hayden, October 23, 1855; NA, RG 57; typescript copy at UW.

30. Hayden to Meek, December 8, 1855; SIA, RU 7062; typescript copy at UW.

31. Hayden to Engelmann, December 21, 1855; NA, RG 57; typescript copy at UW.

32. Hayden to Baird, December 26, 1855; SIA, RU 7002; typescript copy at UW.

33. David Rice Atchison (1807-1886) was a lawyer who served in the Missouri Militia, the Missouri legislature, and the U.S. Senate. Following passage of the Kansas-Nebraska Act he was a leader in the attempt to admit Kansas as a slave state.

34. Edwin Thompson Denig (1812-1858) was a friend and astute observer of the Indians of the Upper Missouri Country. His definitive descriptions of Indian life and culture remained in manuscript form until edited and published by John C. Ewers in 1930. He was employed as a bookkeeper at Fort Union.

35. Hayden to Baird, January 9, 1856; SIA, RU 7002; typescript copy at UW.

Epilogue

1. Warren to Hayden, February 15, 1856; reproduced in Merrill, *The First Hundred Years of American Geology*, p. 117.

Works Cited

Archival Holdings Cited

Spencer Baird Papers, Smithsonian Institution Archives, Smithsonian Institution, Washington, D.C. [SIA]

George Engelmann Papers, Missouri Botanical Garden Archives, St. Louis, Missouri. [MBGA]

Fritiof M. Fryxell Papers, Special Collections, Augustana College Library, Rock Island, Illinois. [ACL]

James Hall Papers, New York State Archives, Albany, New York (Accession No. KW 12835). [NYSA]

Ferdinand V. Hayden Collection, American Heritage Center, University of Wyoming Library, Laramie, Wyoming. [UW]

Hough Collection, Iowa Historical Society, Des Moines, Iowa (Collection H-3244).

Joseph Leidy Collection, Manuscript Archives, Academy of Natural Sciences of Philadelphia (Collection 1-B). [ANSP]

J. Peter Lesley Papers, American Philosophical Society, Philadelphia, Pennsylvania.

George P. Merrill Papers, Library of Congress, Washington, D.C.

National Archives, Record Group 57. [NA]

John Strong Newberry Papers, Western Reserve Historical Society, Cleveland, Ohio.

Oberlin College Archives, Oberlin College, Oberlin, Ohio.

Published Works Cited

Albany [New York] Medical College, Catalog and Circular, Fall Term 1853.

Athearn, Robert G. *Forts of the Upper Missouri.* Englewood Cliffs, NJ: Prentice-Hall, 1967.

Boaden, Ann, and Richard C. Anderson. Memorial to Fritiof Melvin Fryxell. *Geological Society of America Memorials.* Vol. 18, unpaged.

Boller, Henry A. *Among the Indians: Four Years on the Upper Missouri, 1858-1862.* Lincoln: University of Nebraska Press, 1972.

Cassidy, James G. *Ferdinand V. Hayden: Entrepreneur of Science.* Lincoln: University of Nebraska Press, 2000.

Chittenden, Hiram Martin. *The American Fur Trade of the Far West*, 2 vols. New York: F. P. Harper, 1902; reprinted Stanford, CA: Academic Reprints, 1954.

__________. *History of Early Steamboat Navigation on the Missouri River.* Minneapolis: Ross and Haines, 1903; reprinted 1962.

Clarke, John M. *James Hall of Albany, Geologist and Palaeontologist, 1811-1898.* Albany, NY, 1921.

Cope, E. D. "F. V. Hayden [Memorial]." *American Geologist*, Vol. 1 (1888), pp. 110-113.

Coulson, Thomas. *Joseph Henry: His Life and Work.* Princeton: Princeton University Press, 1950.

Culbertson, Thaddeus A. "Journal of an Expedition to the Mauvaises Terres and the Upper Missouri in 1850." Smithsonian Institution, Annual Report 5 (1851), pp. 84-132.

__________. "Journal of an Expedition to the Mauvaises Terres and the Upper Missouri in 1850," ed. John F. McDermott. Washington, D.C.: Bureau of American Ethnography, Bulletin 147, paperback edition (1952).

Dictionary of Scientific Biography. Ed. Charles Coulston Gillespie. New York: Scribners, 1972.

Elliott, Clark A. *Biographical Dictionary of American Science: The Seventeenth through the Nineteenth Centuries.* Westport, CT: Greenwood Press, 1979.

Evans, Noel, and Fritiof M. Fryxell. "Jesse V. Howell (1891-1979)." Tulsa, OK: *American Association of Petroleum Geologists Bulletin*, Vol. 55 (1971), pp. 1885-1887.

Ewers, John C. *The Blackfeet.* Norman: University of Oklahoma Press, 1958.

Finley, Jeanne. "Acrimony in Albany." *New York Archives* (Fall, 2004), pp. 18-20.

Foster, Mike. *Strange Genius: The Life of Ferdinand Vandeveer Hayden.* Niwot, CO: Roberts Rinehart Publishers, 1994.

Fryxell, Fritiof M. "Thomas Moran's Journey to the Tetons in 1879." Rock Island, IL: Augustana Historical Society Publications, No. 2 (1934), pp. 37-46.

______________. "Thomas Moran: An Appreciation." *Sierra Club Bulletin* (San Francisco), June 1935, pp. 56-57.

______________. "Thomas Moran's Journey to the Mount of the Holy Cross in 1874." *Trail and Timberline* (Colorado Mountain Club, Denver), No. 203 (1936), pp. 103-105.

______________. "William H. Jackson: Photographer, Artist, Explorer." *American Annual of Photography* (1939), pp. 1-13.

______________. *Thomas Moran: Explorer in Search of Beauty.* East Hampton, CT: East Hampton Free Library, 1958 [84 pages].

______________. "The Thomas Moran Art Collection of the National Parks." *Yosemite Nature Notes,* Vol. 15 (1958), pp. 57-60.

______________. "Albert Charles Peale (1849-1914), Pioneer Geologist of the Hayden Survey." *Annals of Wyoming,* Vol. 34 (1962), pp. 175-192.

Fryxell, Fritiof M., and Noel Evans. "Memorial to Jesse V. Howell (1891-1971)." Boulder, CO: *Geological Society of America Memorials,* Vol. 3, pp. 109-115.

Gilman, Daniel C. *The Life of James Dwight Dana.* New York: Harper and Brothers, 1899.

Goetzmann, William H. *Army Exploration in the American West, 1803-1863.* New Haven: Yale University Press, 1959.

Hall, James, and F. B. Meek. "Descriptions of New Species of Fossils from the Cretaceous of Nebraska." New York: American Academy of Arts, Memoir, new series 5 (1856), pp. 379-411.

Hanson, Joseph Mills. *The Conquest of the Missouri: Being the Story of the Life and Exploits of Captain Grant Marsh.* Chicago: A.C. McClurg & Co., 1909; reprinted New York: Murray Hill Books, 1946.

Hayden, F. V., et al. *On the Geology and Natural History of the Upper Missouri.* Philadelphia: C. Sherman & Son, 1862.

Hendrickson, Walter Brookfield. *David Dale Owen: Pioneer Geologist of the Middle West.* Indianapolis: Indiana Historical Bureau, 1943.

Herber, Elmer C. "Spencer F. Baird: A World Famous Naturalist from Cumberland County." N. p. [perhaps Carlisle College], 1954 [7 pages].

Hodges, William Romaine. *Carl Wimar.* Galveston, TX: Charles Reymershoffer, 1908 [37 pages].

Kurz, Rudolph F. "Journal of Rudolph Friederich Kurz: An Account of His Experiences among Fur Traders and American Indians on the Mississippi and the Upper Missouri Rivers during the Years 1846 to 1852." Translated by Nyrtis Jarrell. Edited by J. N. B. Hewitt. Washington, D.C.: Bureau of American Ethnology, Bulletin 115 (1937).

Lange, Erwin F. "Dr. John Evans (1812-1861), U. S. Geologist to the Oregon and Washington Territories." *Proceedings of the American Philosophical Society*, Vol. 103 (1959), pp. 476-484.

Larpenteur, Charles. *Forty Years a Fur Trader on the Upper Missouri.* Chicago: The Lakeside Press, 1933.

Leidy, Joseph. *The Ancient Fauna of Nebraska.* Smithsonian Contributions to Knowledge, Vol. 6 (1853) [126 pages].

Lesley, J. Peter, "Obituary Notice of Ferdinand Vandiveer Hayden M.D., Ph.D., L.L.D." American Philosophical Society Miscellaneous Pamphlets, 26 (1888) [61 pages].

Lurie, Edward. *Louis Agassiz: A Life in Science.* Chicago: University of Chicago Press, 1960.

McDonnell, Anne, ed. "The Fort Benton Journal, 1854-1856, and The Fort Sarpy Journal, 1855-1856." *Contributions to the Historical Society of Montana*, Vol. 10 (1940) [327 pages].

Meek, Fielding B., and Ferdinand V. Hayden. "Descriptions of New Species of Gastropoda from the Cretaceous of Nebraska Terr." *Proceedings of the Academy of Natural Sciences of Philadelphia*, Vol. 8 (1856).

Merrill, George P., ed. "Contributions to a History of American State Geological and Natural History Surveys." United States National Museum, Bulletin 109 (1920).

Merrill, George P. *The First Hundred Years of American Geology.* New Haven: Yale University Press, 1924.

Nelson, Clifford M. "Toward a Reliable Geologic Map of the United States." In *Surveying the Record: North American Scientific Exploration to 1930*, ed. Edward C. Carter II. Philadelphia: American Philosophical Society, 1999.

Nelson, Clifford M., and F. M. Fryxell. "The Antebellum Collaboration of Meek and Hayden in Stratigraphy." In *Two Hundred Years of American Geology*, ed. Cecil J. Schneer. Durham, NH: University Press of New England, 1979.

Norwood, William Frederick. *Medical Education in the United States before the Civil War*. Philadelphia: University of Pennsylvania Press, 1944.

Oesher, Paul H. *The Smithsonian Institution*. New York: Praeger Publications, 1970.

O'Harra, Cleophas C. "The White River Badlands." South Dakota School of Mines, Department of Geology, Bulletin No. 13 (1920) [181 pages, 96 plates].

Owen, David Dale. *Report of a Geological Survey of Wisconsin, Iowa, and Minnesota*. Philadelphia: Lippincott, Grambo and Company, 1852.

Palmquist, Peter, and Thomas Kailbourn. *Pioneer Photographers of the Far West: A Biographical Dictionary, 1840-1865*. Stanford: Stanford University Press, 2000.

Partoll, Albert J. "Blackfoot Indian Peace Council." *Frontier and Midland*, Vol. 17 (1937), pp. 199-207.

Prout, Hiram A. "A Description of a Fossil Maxillary Bone of Paleotherium from Near White River." *American Journal of Science*, 2nd Series, Vol. 3 (1847), pp. 248-250.

Rabbitt, Mary C. *Minerals, Lands and Geology for the Common Defence and General Welfare*. Washington, D.C.: U.S. Government Printing Office, 1979.

Seymour, Flora Warren. *Indian Agents of the Old Frontier*. New York: D. Appleton-Century Company, 1941.

Stevens, Hazard. *The Life of Isaac Ingalls Stevens*, 2 vols. Boston: Houghton, Mifflin, 1901.

Stevens, Isaac Ingalls. Correspondence of the Office of Explorations and Surveys concerning Isaac Stevens' Survey of the Northern Route for the Pacific Railway. Washington, D.C.: U. S. Office of Explorations and Surveys, 1947.

Stevenson, John J. "John Strong Newberry." *American Geologist*, Vol. 12 (1893), pp. 1-25.

Taylor, Emerson Gifford. *Gouverneur Kemble Warren*. Boston: Houghton, Mifflin, 1932.

Thomas, Davis, and Karin Ronnefeldt, eds. *People of the First Man: Life Among the Plains Indians in their Final Days of Glory: The First-hand Account of Prince Maximilian's Expedition up the Missouri River, 1833-1834*. New York: E.P. Dutton, 1976.

Thoms, Herbert. *The Doctors Jared of Connecticut: Jared Eliot, Jared Potter, Jared Kirkland*. Hamden, CT: The Shoe String Press, 1958.

Vaughan, Alfred J. "Report of Alfred J. Vaughan," No. 28. In *Annual Report of the Commissioner of Indian Affairs for 1854.* 33rd Congress, 2nd Session, 1854, Senate Executive Document 1, pp. 287-297.

Vaughan, Alfred J. "Report of Alfred J. Vaughan," No. 24. In *Annual Report of the Commissioner of Indian Affairs for 1855.* 34th Congress, 1st Session, 1855, Senate Executive Document 1, pp. 391-398.

Warren, G. K. "Explorations in the Dacota Country in the Year 1855." 34th Congress, 1st Session, 1856, Senate Executive Document 76. [Hayden's report is contained in Appendix E, pp. 66-79; two maps accompany this report.]

White, Charles A. "Memoir of Ferdinand Vandiveer Hayden, 1839 [*sic*]-1887." National Academy of Science, Biographical Memoirs 3 (1895), pp. 395-413.

Wilson, F. T. (with editorial notes by Charles E. DeLand). "Old Fort Pierre and Its Neighbors." Pierre, South Dakota: South Dakota Historical Society, Vol. 1 (1902), pp. 257-379.

Yochelson, Ellis. "Fielding Bradford Meek." *Dictionary of Scientific Biography*, Vol. 9 (1974), pp. 255-256.

Index